SMART WIRELESS SENSOR NETWORKS

SMART WIRELESS SENSOR NETWORKS

Editor

Izaan Ahmed

scitus academics

Smart Wireless Sensor Networks
Edited by **Izaan Ahmed**

ISBN: 978-1-68117-605-5
Library of Congress Control Number: 2016946460

SCITUS Academics LLC,
616, Corporate Way, Suite 2, 4766,
Valley Cottage, NY 10989

www.scitusacademics.com

Notice

Printed in United States of America on Acid Free Paper

PREFACE

The Wireless Sensor Network (WSN) has become one of the most promising technologies in the future. This was made possible by advances in the availability and technology of small, affordable and intelligent sensors, which resulted in economical and rapid deployment of WSNs. However, researchers must tackle a number of challenges in order to facilitate the wide deployment of WSN technology in real-world fields. Facilitated with intelligent computing and wireless communication capabilities, these nodes begin to be intelligent sensors that not only perceive ambient physical parameters, but also have the ability to process information, cooperate with each other in the network. Smart Wireless Sensor Networks discusses various aspects of the design of such smart wireless sensor networks.

The book covers design methodologies, network protocols and algorithms, quality of service management, coverage optimization, time synchronization and sensor network security techniques. These new features also help the sensor nodes to better function both in terms of data acquisition and energy use. Specific application purposes require the operation and design of WSNs that differ from standard networks, such as the website. There are many different areas where sensor networks are applied. For example, forest fires can be identified through sensor networks so that they can be combated at the beginning. Sensor networks can be used to monitor the structural integrity of civil structures through the location of damage in bridges, for example. In addition, they are used to monitor human physiological data in the health sector. They play an excellent role in remote monitoring and demand management, which enables business processes such as real-time pricing.

TABLE OF CONTENTS

1

WSN- AND IOT-BASED SMART HOMES AND THEIR EXTENSION TO SMART BUILDINGS

Hemant Ghayvat, Subhas Mukhopadhyay *, Xiang Gui and Nagender Suryadevara

SEAT, Massey University, Palmerston North 4442, New Zealand

ABSTRACT

Our research approach is to design and develop reliable, efficient, flexible, economical, real-time and realistic wellness sensor networks for smart home systems. The heterogeneous sensor and actuator nodes based on wireless networking technologies are deployed into the home environment. These nodes generate real-time data related to the object usage and movement inside the home, to forecast the wellness of an individual. Here, wellness stands for how efficiently someone stays fit in the home environment and performs his or her daily routine in order to live a long and healthy life. We initiate the research with the development of the smart home approach and implement it in different home conditions (different houses) to monitor the activity of an inhabitant for wellness detection. Additionally, our research extends the smart home system to smart buildings and models the design issues related to the smart building environment; these design issues are linked with system performance and reliability. This research paper also discusses and illustrates the possible mitigation to handle the ISM band interference and attenuation losses without compromising optimum system performance.

KEYWORDS

IOTs; wellness function; behavioral detection; attenuation loss; interference; SNR (signal to noise ratio)

1. INTRODUCTION

Innovations in technology mostly emerge from the needs of human society. The 21st century is the era of prompt advancement in digital technology. Most of this technology is focused on proficiently monitoring and controlling different activities. Everywhere from mega-structure building automation to small smart homes, big industrial assembly machineries to a kid's toy, a college research laboratory to an international space research center, and even health care service at a desk through wireless sensors and networks, wireless sensor networks (WSN) have become fundamental and crucial devices. The significant improvement offered by introducing wireless technology is that it reduces the complexity to harness wired transmission and facilitates the installation of sensors, controllers, and actuators. The cost and installation efforts for a large number of sensors in an urban environment are exponentially reduced by wireless technology innovations. There are different wireless communication mediums (technology) in which a wireless sensor network can be constructed according to respective applications and strengths.

Home automation and monitoring are the dominant applications of WSNs, where a number of heterogeneous sensors are deployed, to determine different activities of inhabitants. Wireless sensors can be operated through batteries as well as plugging into the power supply. The selection of power source depends on the deployment environment and the availability of power for applications such as solar panel based outdoor temperature sensing. In a replaceable battery, the sending power and sampling rate of data decide battery life while the energy harvesting applications use the external energy source, so they are somewhat free from the battery usage method. Wireless sensors and networks are comprised of several nodes prepared through various sensing devices (sensors, controllers, and actuators) and RF chips for wireless communication [1,2,3,4]. A revolutionary development to WSNs is the announcement of IEEE 802.15.4 standard in the year 2003 [5], and it is the first major worldwide standard for WSNs. The limitation of IEEE 802.15.4 standard is that, it only specifies and defines the RF communication with respect to the lower layers PHY and MAC; it provides proposals and does not define the networking techniques for the upper layers. In light of this, ZigBee Alliance and its Mesh network along with IPv6 (in recent years) standardized the protocol known as ZigBee communication protocol (IEEE 802.15 ZigBee protocol). These upper layer improvements offer authentication of network nodes, encryption, and an efficient and modern routing that leads to mesh networking topology. Although mesh topology is quite complex, ZigBee is mostly preferred by WSNs designers with Mesh topology only. IEEE 802.15.4 ZigBee offers tremendous specifications to short range and urban environment wireless sensors networks. IEEE 802.15.4 ZigBee standards operate on the license-free industrial, scientific, medical (ISM) frequency band.

One of the significant developments in WSN-based smart homes is the injection of IOTs (the internet of things). The ubiquitous connectivity and distributed intelligence of the IOTs with wireless sensing technology are becoming the center point of upbeat remote monitoring and control. In an era of IOTs, more devices are linked to the internet as compared to the population—there were over 12.5 billion devices in 2010. Cisco predicts that 25 billion devices are going to come onboard in 2015 and there will be 50 billion by 2020. The internet of things from a smart home perspective can be described as connecting household objects, especially electronic and electrical appliances and sensors and actuators, to the World Wide Web services [6,7]. These objects and devices are smartly and intelligently interlinked to each other to develop new customs and understanding of wireless communication between a person and things, and among things themselves. These customs and understanding create a prospect for devices nearby in the smart home environment to start the interaction and build very different environments. An environment that learns from our daily activity and requirements, such as when we get up in the morning, how much time we sleep, and how long we watch television, can react to an inhabitant's behaviour to improve their health and wealth. For example, an occupant got up late in the morning and because he or she was in hurry to catch the bus to the office, he/she forgot to turn off the electric oven and only noticed this after reaching the office. Now from a remote distance, through an IOT-based smart home system, he or she can control the oven.

The motivations to design and develop the smart home are: independent living; enhancing comfort, efficient use of electricity, and safety and security. The word 'smart home' is preferred for a home environment equipped with advanced technology that allows monitoring and control of its inhabitants, and boosts independent living through wellness forecasting based on behavioral pattern generation and detection. To identify the difficulties and challenges of the prime performance of smart home monitoring, we have to understand recent and ongoing research in this field. A variety of smart home systems for ambient assisted living has been proposed and developed, but there are, in fact, relatively few houses that apply smart technologies. One of the main reasons for this is that the complexity and varied design requirements associated with different domains of homes. These domains are communications, control, entertainment, residential and living spaces. AHRI (Aware Home Research Initiative) at Georgia Institute of Technology [8], CASAS (The Center for Advanced Studies in Adaptive Systems) at Washington State University [9], AgingMo at University of Missouri [10], PlaceLab at MIT [11] and Smart home Lab at Iowa State University [12,13], are monitoring the activity of daily living, enhancing the comfort, and creating context-aware relative scenarios through heterogeneous sensor deployment, and these sensor units include cameras as well. These are academic research projects. From a social point of view, the use of the camera is a direct violation of privacy for monitoring in a smart home environment. At the same time, the video stream by the camera is technically complex. Monitoring and analysis through video stream are simple, but it demands significant storage space on the server. The server designed at home is a local home gateway server, which has limited storage and collection memory. There are many different industrial big building smart home services available; Toyota's dream house at Japan, e-House by McDonough, in New York, House R128 by Intgeniere, W.S at Stuttgart, Germany and Crystal House by Hung, at Taichung, Taiwan, which have targeted sustainability and energy conservation goals, and offer convenience and comfort through the intelligence of the environment [14].

Wearable, implantable, and microsystems that can be deployed over the body area network such as the Apple watch are available nowadays. These devices are worn by the individual or embedded in the home environment to assist someone for health care. Several projects comprising the application of signal assisted wearable devices are ongoing in different research laboratories, SmartShirt [15], Senswear Armband by Body Media Inc [16] and LifeShirt by Vivometrics [17], PROETEX project by CSEM center [18] are wearable devices to measure ECG. Other research has focused on fall detection based on accelerometers using wearable sensors [19,20,21,22,23,24]. An individual suffering from illness can wear these devices as part of a health care strategy, but someone who is fit may not appreciate wearing these devices on their body. Adaptability is a significant concern with wearable sensor devices for physiological parameter measurement.

Other researchers have already implemented smart home systems based on wireless communication technology, but integrating wireless with IOTs has extraordinary potential in smart home applications. Most research of IOT-based smart homes are in the proposal phase, and very few research studies involve practical implementation [25,26,27, 28,29,30,31]. IOT-based smart systems offer us remote access and reactive ambient analysis of home ambience, but they can only perform this when the implementation of IOTs is done through dynamic database management [32,33,34,35]. Additionally, the majority of smart home projects are carried out systematically based on the assumptions from laboratory or test bed controlled environments, which may be entirely different from a real environment where an inhabitant lives regular life regardless of sensor node installation.

The smart home environment demands activity recognition of daily living from raw sensor data, and these raw data sets are complex and irregular to encode into predefined scenarios. Even after encoding this raw data, it is quite difficult to identify deviant behavior because these data sets are on different time measures (sampling rate) and sense modalities. These different time and sensor modalities create problems during equation formulation. It is quite easy to generate behavioral patterns from single event data sets, but it may raise a false alert. In our research, we present a realistic approach to getting the optimal performance from a smart home monitoring system. Realistic application for behavioral pattern generation for ambient-assisted living raises different challenges. To provide a reliable solution, we proposed and implemented a smart home based on an integrated framework to analyze the individual's behavior from the previous data, real-time recently received data and feedback received data. To resolve the issues, we apply pattern detection through pipeline processing with filtering, characteristic construction, activity detection, and smoothing. We attempt to separate the usual routine data from unexpected data, which may put an inhabitant's health and wealth at risk.

The system is proposed and implemented on two levels; hardware and software. At the hardware level, heterogeneous sensors are deployed for multi-activity and multi-events; these wireless sensor nodes are configured with ZigBee Mesh topology and data is received by central coordinator node and collected by local home gateway computer. The software modules are subdivided into different levels, such as data logging, data extraction, and data storage; but their ultimate task is to forecast the change in activity and correlate it with the wellness of inhabitants in real time or near time.

In the second half of the paper, we have extended the smart home approach to big building environments and investigated performance and issues, especially interference and attenuation issues. Smart home monitoring is based on the ambient scenario; when extend our research to smart buildings we are faced with some new problems that we do not notice or consider in a smart home.

We can define the intelligent and smart monitoring as an application of automation with integral systems of accommodation facilities to boost and progress the everyday life of an occupant. In this research, we have designed an XBee series-2 based intelligent monitoring system that operates on the ZigBee protocol and uses the features of IOTs. Our idea is to generate a system-level design approach to formulating design issues and possible solutions. The remaining parts of this paper are organized as follows. We start by presenting the existing system in Section 2 and experimental results and performance issues in Section 3. The extensions of a smart home for big building environments are presented in Section 4. Measurements and investigations from big building environment experiments are presented in Section 5. Lastly, a conclusion with possible solutions to issues is drawn in Section 6.

2. DESCRIPTION OF THE DEVELOPED SMART HOME MONITORING SYSTEM

Smart homes for ambient-assisted living involve the learning patterns of agent behavior from sensor data. Deviation from the regular behavior is an array of the data that either does not follow the anticipated behavior, which is an anomalous behavior, or follows the certain class of predefined irregular behavior. This deviation is an alert for an inhabitant. In our research for behavioral pattern generation, we used an advanced approach that is based on combined deviation over time because forecasting can be reliable only if we consider more than one event. The wellness forecasting model we have designed and proposed is divided into two sections. At the lower section, all the raw sensor data from household appliance usage and movement is delivered to the coordinator attached to the local home gateway and this local server stores the unstructured data sets for further processing and analysis. Raw sensor data collected at the lower information section can only identify which sensor is active or inactive and at what time. To discover the optimal activity of daily living for wellness, these data sets need to be analyzed by the upper section software logic. Intelligent data recognition and classification mechanisms are applied through the software at different levels of data generalization in real time, based on the time and order of the sensor usage.

There are various ZigBee RF modules available in the wireless industry. For our research application, we pick Digi XBee Series-2 RF. The XBee module facilitates various features such as different sampling rate, baud rate and sleep rate associated with two types of operating modes. The first is an application programming interface (API), and the other is application transparent (AT).

2.1. Topology and Device Configuration

Each ZigBee based wireless sensor networks has a ZigBee Coordinator (ZC). There is precisely one coordinator in every single network, and ZC is the device that takes the network establishment responsibility. ZC stores the network information including security keys (such as personal area network identification number). This coordinator is the transceiver RF module that not only receives the data from respective sensor nodes belonging to it, but also looks after remote configuration and fault detection of other associated sensor nodes in the network. In the tree and mesh topologies, the ZigBee routers (ZRs) is applied to extend the network coverage area for wireless communication. In addition to ZC and ZRs, there are two more types of ZigBee devices. The ZigBee device can be configured (programmed) as ZigBee end device (ZED) and ZigBee end device plus router (ZEDR). ZED is usually low power as well as small battery power devices. They transfer their data directly to the parent (ZC), and the parent may be the coordinator or another router node. The ZED cannot relay data from other nodes. The ZRs relay the packet of their neighbor to designated path while the ZEDR node can relay its data and other neighbors' data as well in the network. The ZR and ZEDR nodes both have the same power option as ZED. The ZigBee nodes are scattered at different distances in the urban area of the home environment, and data is sent over the network to the coordinator.

2.2. Forming a Network

ZigBee based Digi XBee Series 2 is used as RF module in our smart and intelligent home monitoring system. The system contains heterogeneous sensors, which include PIR, temperature, force and electrical and electronic appliances usage monitoring sensors. The ZC has the authority to select a channel, PAN ID (16-bit personal area network unique identification number that only belongs to particular ZigBee WSN), security policy, and stack profile for the network. Each time to establish communication, ZC starts with the proper channel and executes a series of scans (energy scan or PAN scan) to realize any RF activity on various channels of ZED, ZR and ZEDR. This energy scan can only register the wireless sensor nodes (ZED, ZR & ZEDR) which are registered with corresponding PAN ID to the ZC.

2.3. Channel Selection and Operation Mode

It is the duty of ZC to select the best channel for network operation. To do this, it performs energy scans on every possible frequency channel (ZigBee 16 channels) to detect the strongest available channel. Channels at excessive energy levels are discarded from potential channel lists to start on. The type of operation mode depends on the individual's requirements. It can be configured by the RF module through XCTU software [36].

2.4. Present System Description

The present home monitoring system has been continuously running since May 2013 without any significant complexity and maintenance, except minor problems related to sensor physical damage and power supply. Figure 1 represents the image of the real smart home where a monitoring system is running, and inhabitants are living their ordinary daily life. This house was built in 1938, so it is an old home, but it has been converted to a smart home with the help of sensing technology. The layout of the smart home with sensor location and placement is shown in Figure 2. This figure displays the genuine home environment where one stays alone and has smart sensing placement to improve wellness of individual, *i.e.*, Ambient Assisted Living (AAL), and it shows different sensing units for monitoring of ambient parameters and different domestic objects. As we can see in the present system, it contains a network of heterogeneous sensors, These heterogeneous sensors are temperature, force, PIR and electronic and electrical (E & E appliances sensing) appliances monitoring units; Brief description of each sensing unit are given as follows:

Figure 1. Image of real house where smart home monitoring and a control system are installed.

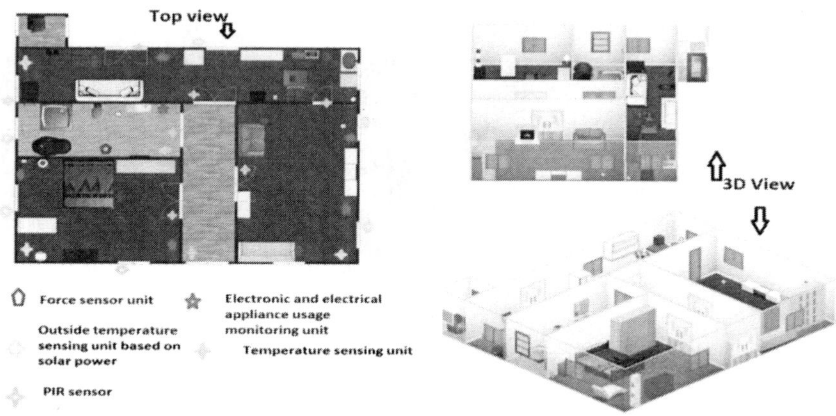

Figure 2. Layout structure of the house with sensor deployment and household objects.

2.4.1. Electronics and Electrical Appliance Monitoring Unit

This power usage monitoring and control unit, contains transformer and other circuit components. The transformer block comprises of voltage and current transformers. The step down voltage transformer is used to transform the mains from 220 V to 10 V signal, and the current transformer ASM010 is used to link the current in the line wire to the load through the current transformer circuit. For signal amplification, operational amplifier LM324 is applied associated with other components such as a rectifier, capacitors and gain resistors of specific parameters. The analog sensor signal output is supplied to the analog channel of radio communication chip for wireless transmission. Figure 3a,b presents the designed unit, and Figure 3c–h shows the utilization of the unit with different household appliances. This data is helpful to analyze the appliance usage pattern [37].

2.4.2. Force Sensing Unit

The flexiforce sensor is a Piezoresistive sensor A301 used to calculate the amount of pressure given on any object. When any pressure is given to the sensor, the resistance of the sensor decreases. As a result, the output voltage increases. The range of resistance varies in force sensor found in different types of sensors. The pressure is ideally given in the central circular part of the sensor on both sides. The sensor is kept beneath the objects to sleep and sit upon. The sensor is connected to a conditioning circuit with the 9 V power supply. Figure 3i shows the force sensing unit and (j) to (l) display the deployment of this unit to measure force. This force measurement is linked and analyzed to identify the occupancy pattern [38,39].

2.4.3. Contact Sensing Unit for Domestic Objects

For the purpose of domestic objects usage monitoring such as self-grooming table and office desk, we fabricated wireless contact sensing systems and connected them. Figure 3m,n shows the fabricated contact sensing unit connected to a grooming table to identify the frequency of usage, and these objects usage are monitored at local home gateway server by ON/OFF values [39].

2.4.4. Temperature Monitoring Unit

To design temperature sensing unit the LM35 IC is connected to conditioning circuit for ambient temperature monitoring.Figure 4a,b represents the outdoor temperature monitoring and Figure 4c shows the indoor room temperature monitoring.

Figure 3. (**a**,**b**) represent the electronic and electrical appliance usage monitoring and control unit and (**c**–**h**) represent the monitoring and control of household appliances through this unit; (**i**) represents the force sensing unit and (**j**–**l**) deployment of the unit for different sitting as well as sleeping objects; (**m**,**n**) represent the contact sensing unit arrangement [39].

2.4.5. Movement Monitoring Unit

The passive infra-red (PIR) movement monitoring unit is designed to detect the motion within the coverage range of the sensing system. This PIR sensing unit is compact, power-efficient, flexible and durable. These sensing units are also known as "IR motion detector". It operates on 5 V–12 V supply. These are binary mode sensors, and they are interfaced with RF XBee modules. Figure 4d shows the fabricated movement monitoring unit deployed in the smart home at the door.

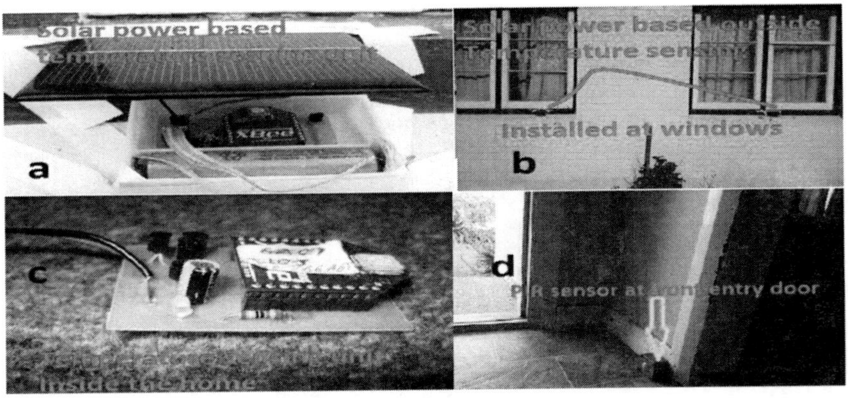

Figure 4. (**a**,**b**) represent the outside temperature monitoring unit based on renewable solar energy; (**c**) inside temperature monitoring unit and (**d**) PIR sensor for activity monitoring.

The sensors are scattered in such a manner so that the systems get every essential value from every corner of the ambient environment that is applicable to wellness determination of an individual who stays there. Figure 3 and Figure 4 show the sensor units' placement in the smart home. These sensing units transmit their data to RF module XBee series-2 end-device/router to XBee Coordinator connected (XBee coordinator connected to the laptop through USB connection) to home gateway server laptop. Figure 5 represents the local home gateway server laptop located in the smart home. The data from local MySQL server is raw data, from this raw data information is extracted through software. The software is designed according to the individual requirement of data extraction. The Visual Studio C# graphical user interface is designed to upload raw data to MySQL server. From the local server, that data will be analyzed and uploaded to the website with the help of the internet [40,41].

Figure 6 represents the movements of a subject on a particular day. The collected passive infrared motion sensor data files were processed and analyzed by sensor activity pattern generation software. The aim of this process was to track the movement of an elderly person and infer the top sequence of movement sensing unit at a particular time (a day or hour of the day). We had restricted the number of movement sensing units, and it was a beneficial approach to identifying the physical location of the person in real-time. We had placed the movement sensing unit exactly at the entry or exit door so that we could quickly identify the physical location of a subject.

Figure 5. Local home gateway server computer.

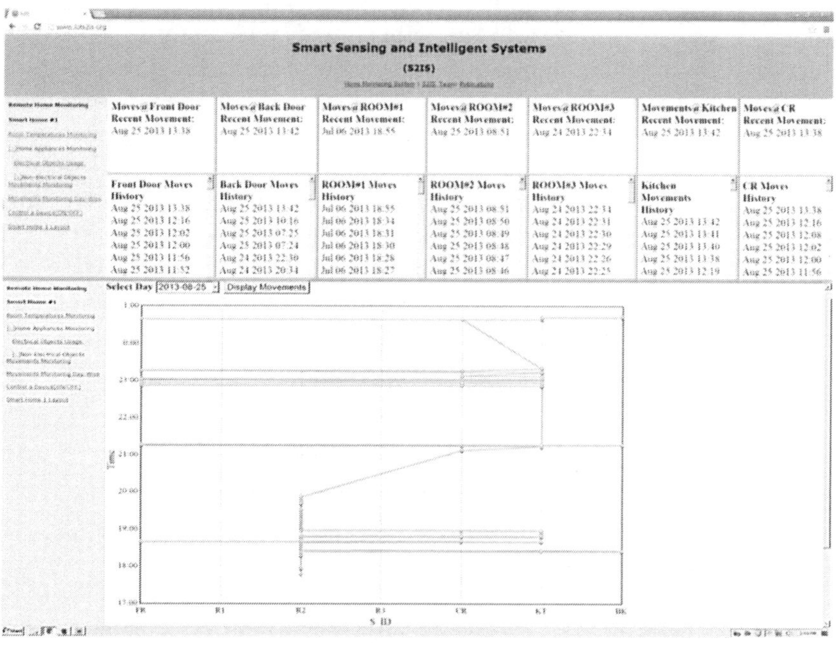

Figure 6. Online smart home monitoring website [39].

2.5. Design of ADLs' Recognition System

The ADLs' recognition system design comprises the native element of the appropriate statistics of the ADLs.

Sensor Event Level 0: This level covers a variety of sensing units deployed in the AAL environment. These are primarily used to produce vital data to the superior level. It transports the heterogeneous sensor data related to the subject statistics and directs the data to the contextual recognition level for classification.

Context Recognition Level 1: This level takes out the information from raw data and, depending on the AAL set-up's principal values of location (S), time (T) and context (C), the contingent information is derived from the identification of Basic ADLs.

ADLs Recognition Level 2: Classification for the Basic-ADLs will be achieved depending on the contextual information and the status of the sensor stream.

At the lower section Level 0, all the raw sensor data from household appliance usage and movement is delivered to the coordinator attached to local home gateway and this local server stores the unstructured data sets for further processing and analysis. Raw sensor data collected at the lower information section can only identify which sensor is active or inactive and at what time. To discover the activity of daily living for wellness, these data sets need to be analyzed by the upper section software logic. The context (Level 1) is intended for recognizing the context at which the sensor value is generated. This level extracts the information from the sensor data and depending on the AAL set-up is the basic values of location, time and context, the situational information are derived from the identification of the respective ADL. The ADL (Level 2) is meant for recognizing the activity based on the context and sensor value, and the labeling for the basic ADL recognition will be performed at this level. Figure 7 shows the recognition of ADLs. It involves a Sensor Event Level, Context Recognition Level, and ADL Recognition Level [39].

2.6. Wellness Functions

Wellness functions are introduced to provide a mathematical expression and to ascertain the wellness of an individual under the ambient-assisted real-time monitoring environment. The two wellness functions β_1 and β_2 define the wellness of an inhabitant based on the usage of household appliances. The rationale for the wellness functions is to determine "how well" the inhabitant is using daily objects. The first function β_1 is obtained from the non-usage as well as the inactive duration of the appliances. While the second function β_2 is generated from the over-usage of few specific appliances. The wellness index shows the behavior of the person with respect to the daily object usages in real-time.

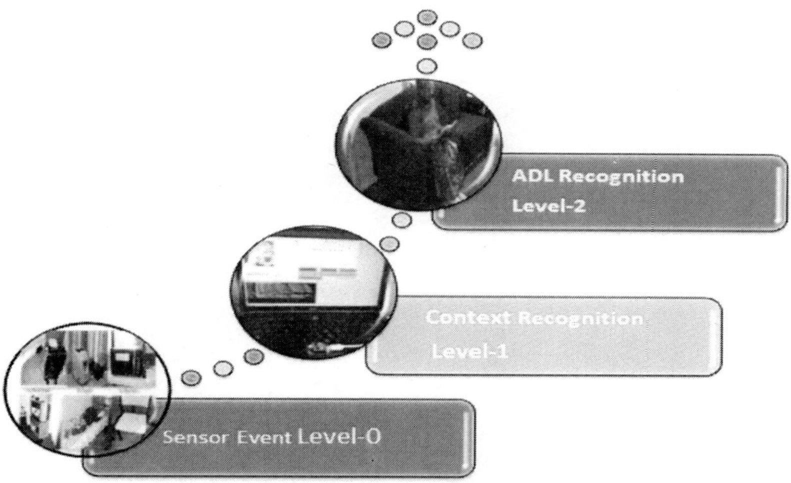

Figure 7. Activity recognition divided into three different levels [39].

Figure 8 shows the functional description of the wellness-based smart home monitoring [37,38,39,40,41]. The smart home systems have been developed to observe the wellness of elderly people living independently in their own at home. The smart home system developed is capable of simultaneously monitoring the general physical activities of an inhabitant as well as physiological and ambient entities. It is a multi-model, unobtrusive, non-invasive novel sensing system deployed at central locations in the home environment. The continuous in-home monitoring can be achieved with single local home gateway server computer. The developed analysis and decision-making algorithm software modules execute in a Windows software working environment. Through the internet connection, we can access the wellness information from a remote location.

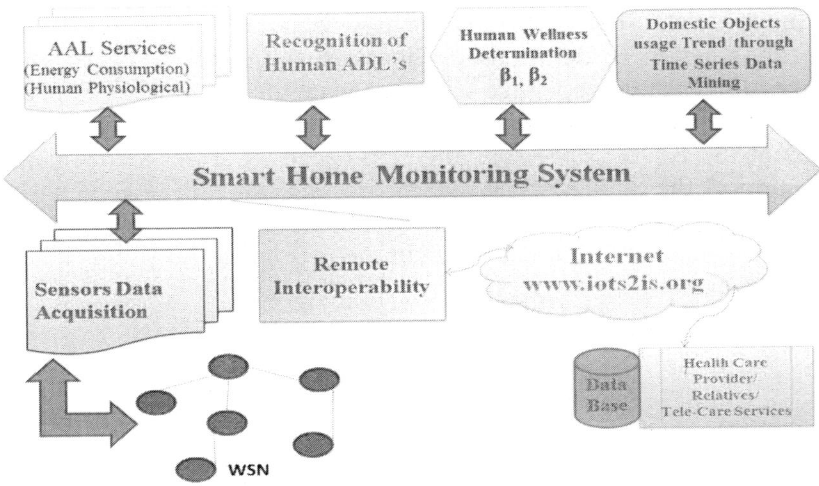

Figure 8. Functional description of the developed smart home monitoring system [39].

Wellness function #1, designated as (β_1) is defined by the following equation:

$$\beta_1 = e^{-t/T} \tag{1}$$

where: β1,New = Wellness index of the elderly person based on the measurement of inactive duration of household objects; t = Time of Inactive duration of all appliances, *i.e.*, duration time no objects are used; T = Maximum inactive duration when no objects were used in the past.

Wellness function #2, designated as (β_2) is defined by the following equation

$$\beta_2 = e^{\frac{(T_n - T_a)}{T_n}} \tag{2}$$

where: β_2, new = Wellness function of the elderly person based on excess usage measurement of household object; T_a = Actual (current) usage duration of the household object; T_n= Maximum usage duration use of household object in normal situation of the past [39].

Seasonal Decomposition

The seasonal (cyclic) decomposition by Brockwell and Davis in 2001 was developed for predicting the near future values. It has been applied as a primary tool to analyze trends. It is also useful for determining seasonal patterns, which improves the forecasting process. Trend component can assess using the principle of moving average.

The initial Exponential Moving Average considered in the analysis is given by the equation below:

$$MA_{t+1} = \alpha X_t + (1 - \alpha)\, MA_t \tag{3}$$

where:

MA_{t+1}—Moving average prediction;

MA_t—Previous Moving average;

α—Smoothing Constant;

X_t—Observed quantity at time 't'.

The smoothing constant ('α'), is derived from the number of sensor observations, and these observations contain the initial value from the start of the system to the recently observed value. The essential features of trend and seasonality are described in a time series by its degree. After assessing the internal components such as trend and seasonality of a time series, errors' estimation is extracted by the de-trending process. Smoothed Trend Curve (STC) for various household usage durations is derived by applying Equation (3).

The seasonal decomposition is suitable for data revealing a cyclical pattern as well as a trend. In the present research task, a one-week activity duration series is considered as one cycle or season to recognize the weekly activity pattern of an individual. It has also categorized the periodic components in the historical data and used them in a forecasting model [39].

Driving Trend using Modified Double Exponential Smoothing Process

To smooth the trends, a modified double exponent smoothing strategy (Brockwell strategy) is applied for forecasting. The advantage of this strategy is to minimize the mean deviation and capture the local (latest seasonal) trend of the series.

$$\tau = T = \delta(L_t - L_{t-1}) + (1 - \delta)T_{t-1} \quad (4)$$

$$L_t = \alpha(x_t - S_{t-s}) + (1 - \alpha)(L_{t-1} + T_{t-1}) \quad (5)$$

$$S_t = \gamma(x_t - L_t) + (1 - \gamma)S_{t-s} \quad (6)$$

where

T_t: trend (or slope) of the entire duration, L_t: local level seasonal slope, S_t: change in seasonal factor, x_t is the observation at the current time, s is the number of periods in one cycle (*i.e.*, s = 7 in our case), α, δ, γ is the smoothing parameters ranging from 0 to 1, selected by minimizing mean square errors.

Starting values are: $L_t = (1/s)(x_1 + x_2 + x_3 + \ldots x_s)$; $T_t = (1/s)((x_{s+1} - x_1)/s + (x_{s+2} - x_2)/s + \ldots (x_{2s} - x_s)/s)$; $S_t = x_k - L_s$, where k = 1, 2….s.

A forecast of the activity duration is extrapolated by using the seasonal pattern in Equations (3)–(5):

$$F_{t+m} = L_t + T_{tm} + S_{t-s+m} \quad (7)$$

m is the required forecast period [39].

3. EXPERIMENT AND ANALYSIS OF SMART HOME

The wellness trends and pattern generation for the activity of daily living are based on the real system running continuously in the real smart home since May 2013. The sensor data is collected into local home gateway computer and analyzed through the wellness function's mathematical formula in Equations (1)–(7).

In the forecasting process, the most appropriate fitted curve is computed by adding smoothed trend curve and seasonally adjusted factors. For illustration, the non-electrical appliances usage durations and their corresponding trends are considered in the forecasting process. This would reveal the precise behavior of an individual in utilizing the household appliances. Some of the electrical appliances such as water kettle, microwave, and laundry machines are pre-programmed and auto control; hence, they may not appropriately run the forecasting method. The non-electrical appliances usage duration and their corresponding trends are plotted. Figure 9 shows bed usage activity durations and its corresponding trend for eight weeks at the smart home where an inhabitant lives their regular life (green color: trend; blue color: actual observations) [37,38,39,40,41,42]. The trend we generated from our wellness model fitted to the actual observations. With these trends, we can predict the bed usage of the next couple of weeks to generate wellness information.

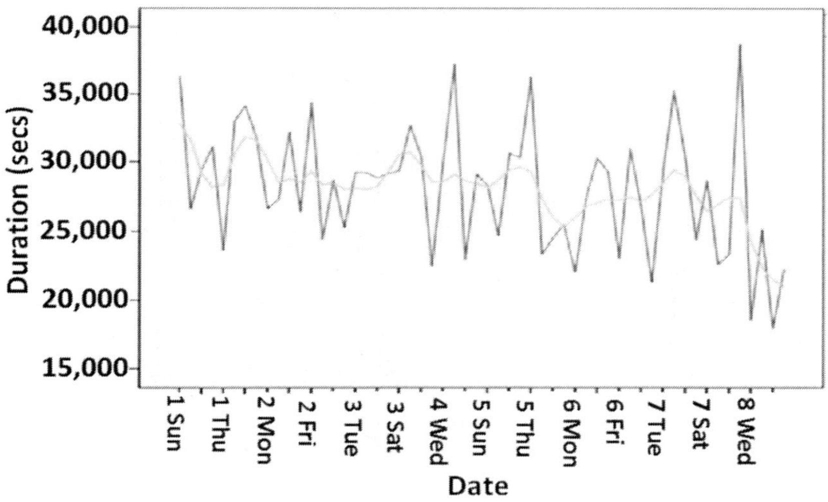

Figure 9. Bed usage durations and trend.

Figure 10 shows the toilet usage activity durations and the corresponding trend for 11 weeks at a smart home where an inhabitant lives (green color: trend; blue color: actual observations). The trend we had generated from our wellness model was close enough to the actual observations.

Figure 11 shows the dining chair usage activity durations and the corresponding trend for 11 weeks at an elderly person's house where he lives alone. The trend we had generated from our wellness model was close enough to the actual observations.

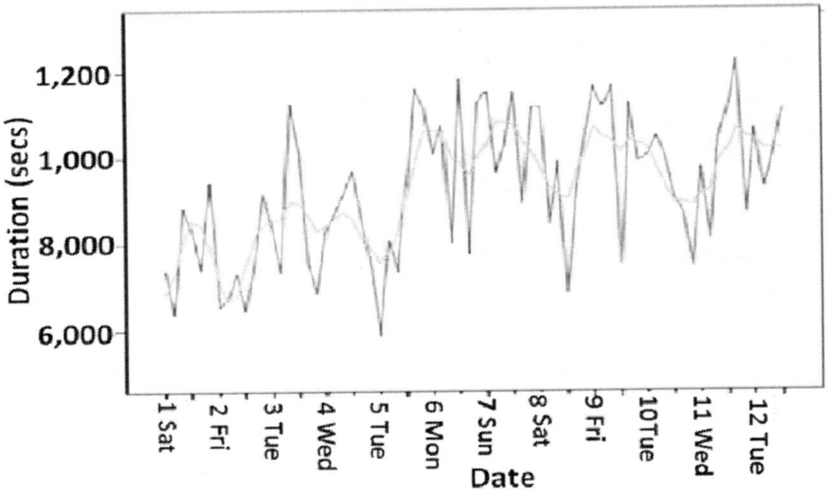

Figure 10. Toilet usage durations and trend.

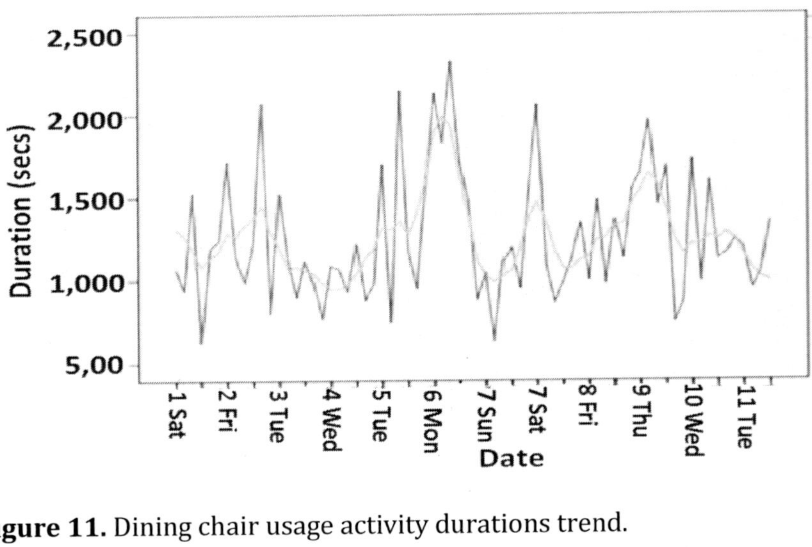

Figure 11. Dining chair usage activity durations trend.

A snapshot of the ninth week's (Friday) estimated values based on the recorded eight weeks is given in Table 1. Considering the statistical inference of 95% confidence interval, the residuals in the fitted (predicted) curve is computed by twice the standard deviation. Accordingly, the forecast ranges with maximum and minimum durations are computed according to Equations (1)–(7).

Table 1. Wellness function indices of household appliances and forecast of the ADLs [39].

SUB	Activity	Sensor_ID	β_1	β_2	Forecasting for 9th Weeks (Friday)					Actual-Duration (s)	Status
					Max-Time (s)	Min-Time (s)	α	δ	Γ		
	Sleeping	Bed		0.829	28,456	20,405	0.120	0.170	0.650	22,505	Regular
	Dining	Chair		0.865	1208	845	0.020	0.110	0.210	1150	Regular
#1	Toilet	Toilet	0.715	0.785	1444	1028	0.140	0.0900	0.350	1353	Regular
	Relax	Couch		0.889	1457	840	0.150	0.130	0.540	1104	Regular
	Watching TV	TV		0.915	2806	2205	0.030	0.140	0.100	2608	Regular
	Sleeping	Bed		0.727	30,258	23,450	0.200	0.180	0.480	27,268	Regular
	Dining	Chair		0.504	14,545	10,425	0.140	0.170	0.300	15,580	Irregular
#2	Toilet	Toilet	0.829	0.576	1838	1426	0.160	0.130	0.200	1718	Regular
	Relax	Couch		0.614	2018	1487	0.150	0.040	0.250	1645	Regular
	Watching TV	TV		0.813	3804	2807	0.020	0.050	0.060	3045	Regular
	Sleeping	Bed		0.816	27,545	21,408	0.030	0.560	0.605	26,258	Regular
	Dining	Chair		0.713	16,250	12,350	0.400	0.600	0.010	15,145	Regular
#3	Toilet	Toilet	0.604	0.883	1628	1245	0.300	0.450	0.500	1465	Regular
	Relax	Couch		0.615	1845	1423	0.205	0.300	0.150	1628	Regular
	Watching TV	TV		0.715	4055	3605	0.100	0.650	0.750	3810	Regular
	Sleeping	Bed		0.445	28,235	22,035	0.205	0.600	0.700	29,701	Irregular
	Dining	Chair		0.914	1340	950	0.400	0.250	0.010	1205	Regular
#4	Toilet	Toilet	0.758	0.818	1123	885	0.100	0.200	0.650	1060	Regular
	Relax	Couch		0.756	1630	1245	0.300	0.200	0.250	1420	Regular
	Watching TV	TV		0.828	4838	4210	0.040	0.100	0.200	4506	Regular

It was observed that two instances of irregularity at different subject houses were rightly predicted. These were related to the over-usage of the appliances. In reality, the subject was using a chair for a longer time because he was sitting and talking with a guest on that day. In another instance, the duration of bed use shows an over-usage because it was occupied by the elderly person for a long duration as he was unwell. The forecasting procedure has indicated the active durations of the bed and chair were outside the forecast ranges. Accordingly, the behavioral detection process set the status of the corresponding activities as irregular.

4. EXTENSION TO SMART BUILDING

After successful implementation of the smart home, we aim to design and discover the issues related to the smart home monitoring system for building apartment where many people live individually. With the existing methodology of the smart home, we installed the sensing system in a big building apartment, although presently we are covering just one floor that has 35 rooms. The heterogeneous sensing systems scattered across the building with layout is shown in Figure 12.

Figure 13 presents the overview of building monitoring system setup. The home gateway system that is connected to the coordinator receives the data from heterogeneous sensors scattered in the home environment under various obstructions.Figure 14 shows the smart building experimental setup to evaluate the attenuation loss effect on WSNs.

The optimum performance in service is only possible with the best reliability. So, for the best reliability user data should be transmitted, received and analyzed within acceptable and defined time duration, in near real time with best precision value and least error.

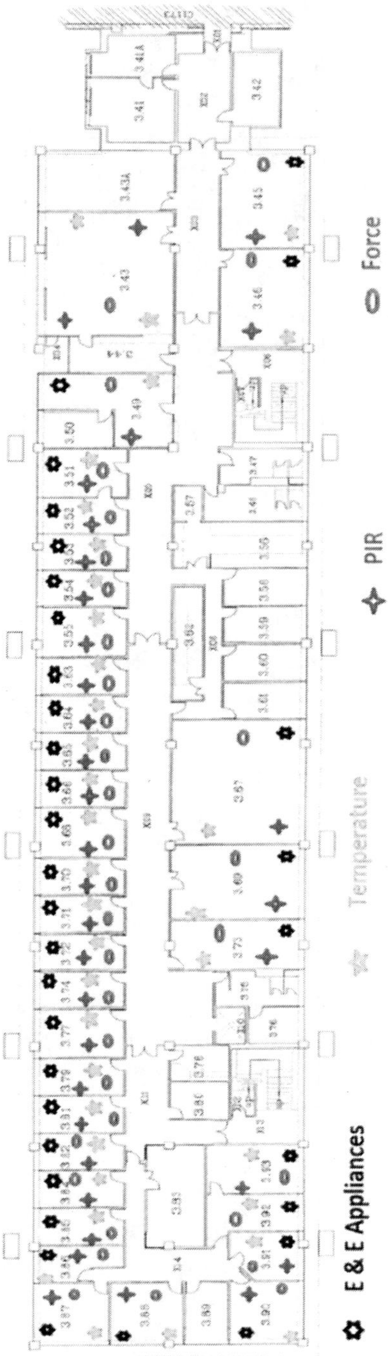

Figure 12. Building layout representing sensors' placement.

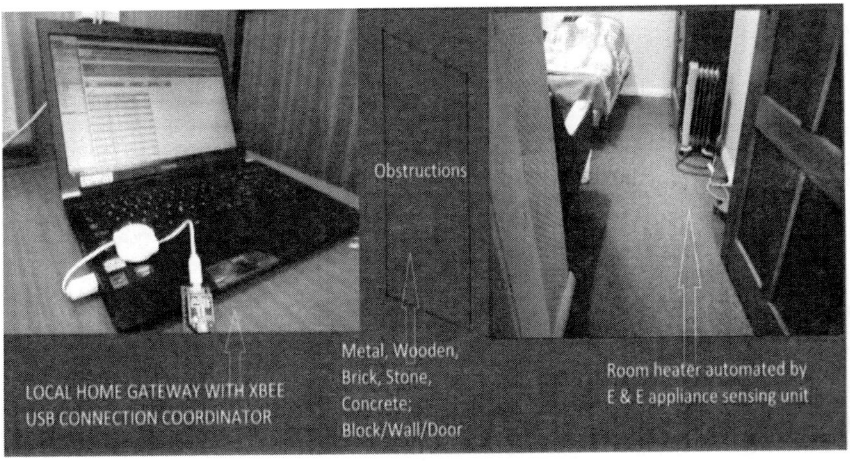

Figure 13. Monitoring system with various types of obstacles.

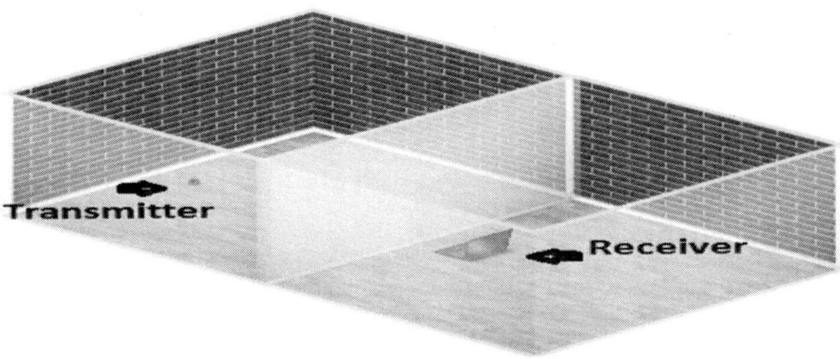

Figure 14. Experimental setup to evaluate the attenuation caused by different building materials in the smart building environment.

In WSN-based smart building, explaining the reliability with the system performance perspective is not straightforward, because in the radio communication we find a number of variables that have a negative impact on reliability. The most common explanation of reliability is data reliability; data transmission, reception, measured error and delay. The data transmission and reception are the function of the communication medium between sensor nodes. The quality of this radio communication link enhances the chances of reliable data delivery. Although the design consideration of the sensor node also decides the performance, they are the sources of data, and if data from the source is corrupted then the system cannot achieve the reasonable accuracy. Most of the time, even with good RF link quality and best sensor node design consideration we get less accuracy. The reason of this is improper routing selection and obstruction in the home environment. The urban environment is full of obstacles, so it causes loss of data. Improper routing and topology increase this data loss more, such as when an end device node does not find the nearby router node and that data would be lost. The signal transmission takes place with the speed of light, so the delay is very small or negligible, although we need to consider this latency because in smart home designing we are focusing on real-time intelligence. The combined delay at the sink will be good enough to affect the near real-time streaming.

We can better understand this by the packet reliability terms of ZigBee based WSNs. Packet delivery ratio (PDR), packet success rate (PSR), packet loss rate (PLR), packet error rate (PER), received signal strength, Signal to noise ratio and received packet delay are some of the parameters that define system reliability and performance. In order to evaluate these parameters in IEEE 802.15.4 ZigBee based wireless sensor and networks, the smart building setup is developed and implemented in real-time to get control and monitoring applications without any time delay. The results show that the distance, deployment environment and positioning of sensor nodes are essential parameters that decide the reliability of wireless sensor and networks.

To understand these results, we have to formulate and understand the packet reliability parameters.

Packet delivery parameters

The packet sent from the source node must be received at sink node without any distortion and manipulation that affect the accuracy. This is usually determined by the number of messages sent from the source and received at the destination. PDR, PSR, PER, and PLR are closely linked to the packet reliability and system performance, and they represent the packet accuracy at a different level.

- PDR: Packet delivery ratio is the number of packets received at coordinator to number of packets sent from the transmitter; it is represented in percentage.

$$PDR(\%) = \frac{N_r}{N_s} \times 100 \quad (8)$$

Nr = Total number of packet received by a coordinator from an end device.

Ns = Total number of the packet sent by the end device.

The higher the PDR value, the better the system performance and packet reliability.

- PSR: Packet success rate is the number of packets successfully received without any error to number of packets received at coordinator; it is represented in percentage.

$$PSR(\%) = \frac{N_{we}}{N_r} \times 100 \tag{9}$$

Nr = Total number of packet received by a coordinator from an end device.

Nwe = Total number of packet received by coordinator without error from an end device.

The higher the PSR value, the better the system performance and packet reliability.

- PER: Packet error rate is the function of number of packets successfully received and the number of packets received at coordinator; it is represented in percentage.

$$PER(\%) = \frac{N_r - N_{we}}{N_r} \times 100 \tag{10}$$

Nr = Total number of packet received by a coordinator from an end device.

Nwe = Total number of packet received by coordinator without error from an end device.

The lower the PER value, the better the system performance and packet reliability.

- PLR: Packet loss rate is the function of number of packets received at coordinator and number of the packet sent from the transmitter; it is represented in percentage.

$$\mathrm{PLR}(\%) = \frac{N_s - N_r}{N_s} \times 100 \tag{11}$$

Nr = Total number of packet received by a coordinator from an end device.

Ns = Total number of the packet sent by the end device.

The lower PLR value offers the better system performance and packet reliability.

We recognized that other than the distance the location of the end-device in the urban environment affects a lot the performance of the system. The quality of the link is highly degraded by urban environmental obstacles such as door, wall, chair *etc.*, followed by ISM band interference and traffic load.

One of the key issues that affect the reliability and performance of WSN-based smart home is intra-network interference that affects packet delivery. Electromagnetic interference is the disruption that upsets desired node signal processing through the electromagnetic radiation emitted from an external source. ZigBee based sensor nodes operate in the 2.4 GHz ISM spectrum. This intra-network interference becomes more critical in unregulated free ISM band of the frequency spectrum. The disturbance sources in this frequency band include:

- Bluetooth (IEEE 802.15.1)
- Wireless USB version 2 (IEEE 802.15.3)
- Wi-Fi (IEEE 802.11)
- Microwave ovens
- Other sources, like some cordless phones and RF motion detectors [43].

We used MetaGeek Spectrum Analysis Wi-Spy DBx & Chanalyzer 5 device as a frequency spectrum analyzer, and configured it for a particular application requirement [43,44]. We are using density and waterfall graphs to model the interference and loss caused by other RF device in ISM band; these graphs represent the RSSI values at different ZigBee channels.

The density view plot represents the signal activities in the selected spectrum to recognize the devices; it shows the signal intensity and how often these devices are transmitting. The height of the graph shows the amplitude of the signal, and the color in the plot represents how often the signals are occurring. The more concentrated the color, the more often the frequency is in use; this is called utilization. A blue spike or profile shows a short signal, like a clap. A red spike or shape shows a long, unbroken signal, like an air horn. Colors and their significance are as follows: blue—less than 10% utilization; green—20% utilization; yellow—40% utilization; red—over 50% utilization.

The waterfall view graphs represent amplitude over time for all frequencies in the selected band; much like a seismometer graphs earthquakes. This view is useful for watching the spectrum over time. Unlike the density view, which uses color by utilization, the intensity of the color in the waterfall view shows amplitude. Blue indicates low-amplitude signals while red indicates high-amplitude signals.

Figure 15 shows the interference caused by other wireless technologies to smart home monitoring ZigBee channel. Our ZigBee based WSNs are functioning at channel 2.425 GHz. The sensor nodes are deployed in Mesh topology into a smart building with spacing between nodes is maximum up to 7 m and the Wi-Fi, Bluetooth and microwave sources are placed at 2 m distance from receiver (coordinator). Figure 15a shows the density and waterfall view of XBee smart building system that is operating at frequency 2.425 MHz under least interference condition. Figure 15b shows the Bluetooth functioning over the same frequency 2.425 MHz, which degraded the XBee RF link quality. Figure 15c shows the Wi-Fi operation that affected the XBee operation badly. Figure 15d shows that microwave oven signals are dissipated across the whole ZigBee spectrum. We can better understand this interference effect by the packet reliability parameters shown in Table 2. The packet reliability metrics are most affected by the microwave oven followed by Wi-Fi, and then Bluetooth.

In the smart home monitoring system, most of the time the radio signal does not find the line of sight communication, and it has to travel through multiple obstacles to reach the receiver end. In real conditions, a radio signal usually comes across different kinds of objects in its transmission path and suffers from attenuation, relying on the absorption characteristics of the obstructions. These objects are of different types—such as mobile and static—and absorb the desired RF signal power and attenuate it. In the same way to the free space propagation loss, higher frequencies cause greater losses by attenuation [36,45].

Table 2. Packet reliability parameters affected by different types of ISM band interference sources.

Title	PDR %	PLR %	PSR %	PER %
(a) Least Interference	100	0	99.56	0.44
(b) Bluetooth Interference	99.50	0.50	99.31	0.69
(c) Wi-Fi Interference	97.85	2.15	99.05	0.95
(d) Microwave oven Interference	90	10	96	4

The experimental setup arranged for this was a realistic, smart building environment as already represented in Figure 13and Figure 14 above. In the building environment, we usually find different architectures and various obstructions such as chairs, tables, and other household equipment, but we chose an empty room as well as a hall where we did not find any household stuffs to get a precise measurement of the attenuation caused by the material. Although in the realistic scenario there are different objects and household equipment, these objects add extra attenuation and multipath propagation that affects the measurement. This setup is made to get the fully controlled environment to examine the attenuation effect on system performance. The transmitter and receiver are separated by a particular type of material and the distance to evaluate attenuation loss was defined.

Initially, with a line of sight path, the RSSI values were recorded and compared with an individual type of obstruction. The attenuation loss by different material types is presented in Figue 16. While the plywood wall does not cause significant attenuation loss, the steel panels used as interior wall in the building causes major attenuation loss.

Human bodies are made of about 70% water; they attenuate RF signals. The attenuation caused by the human body in ISM band is about 3 dB and in the 5 GHz band it is about 5 dB [36,45].

So, to evaluate how a densely populated area affects the system performance, we considered two scenarios, the first was the regular days at home and the second during a party with a group of people celebrating. As shown in Figure 17, the packet reliability in the regular population was slightly better than densely-populated party condition.

From Figure 17, we can easily identify that there is a marginal improvement in system performance and packet reliability but even this matters. In smart home applications, each and every packet has importance, so even this marginal improvement plays a significant role during any emergency assistance support request.

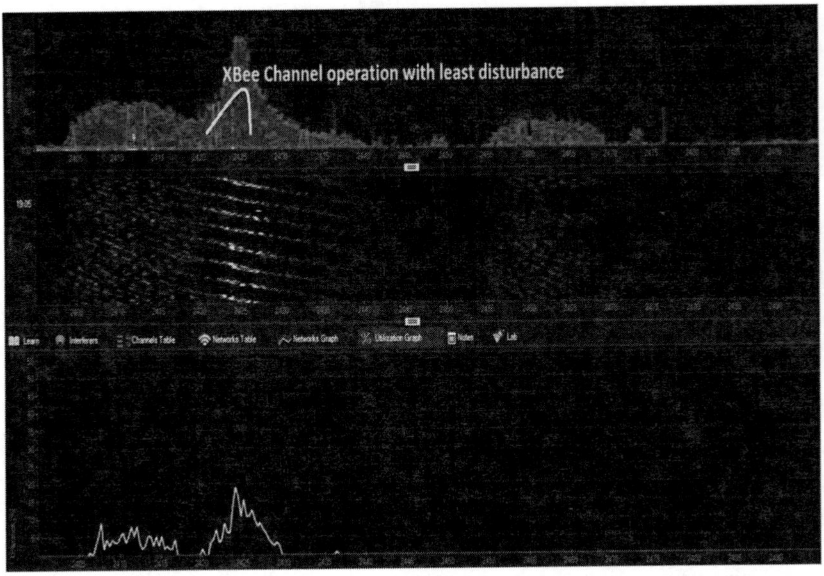

(a)

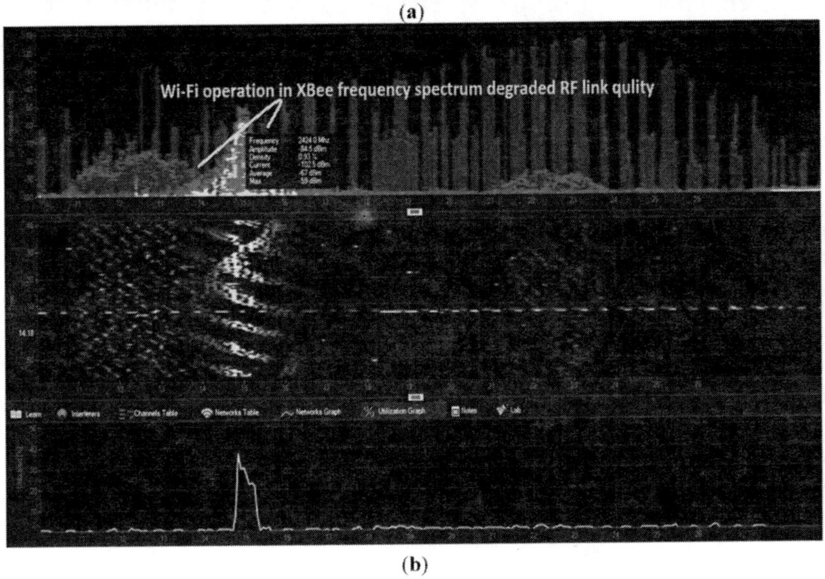

(b)

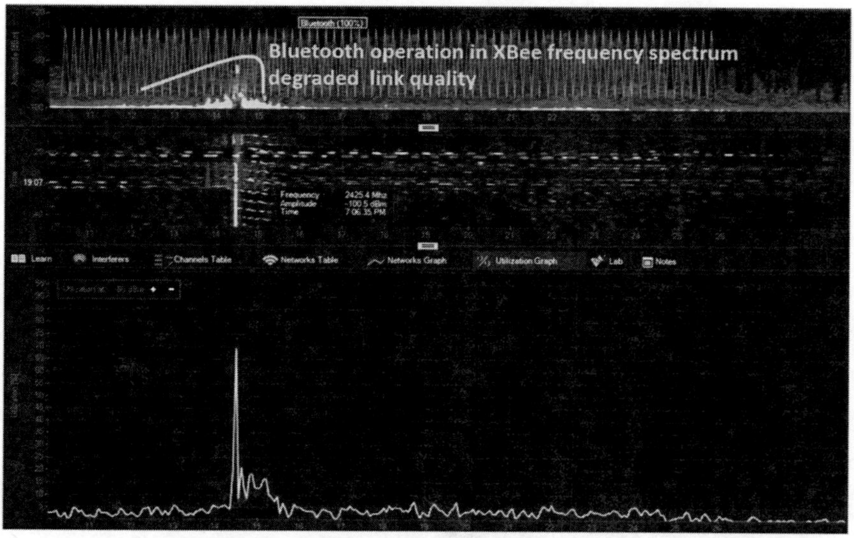

(c)

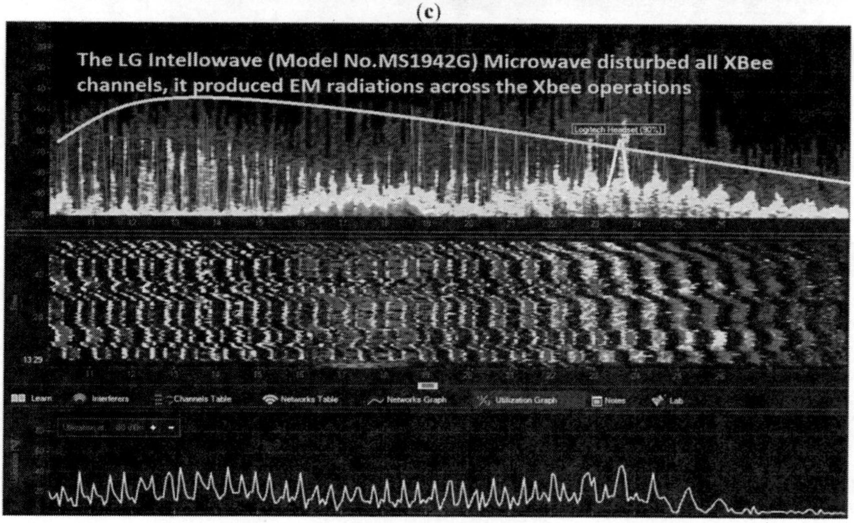

(d)

Figure 15. (**a**) shows the that the XBee smart building system is operating at frequency 2.425 MHz; (**b**) shows the Bluetooth functioning over the same frequency 2.425 MHz, which degraded the XBee RF link quality; (**c**) shows the Wi-Fi operation that affected the XBee operation badly; (**d**) Microwave oven distributed all ZigBee channels, and the microwave signal are dissipated across the whole ZigBee spectrum.

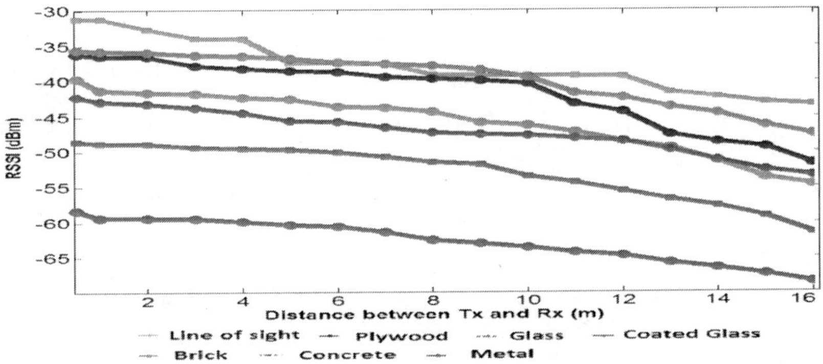

Figure 16. Attenuation loss caused by different materials in the smart building, measured between transmitter and receiver.

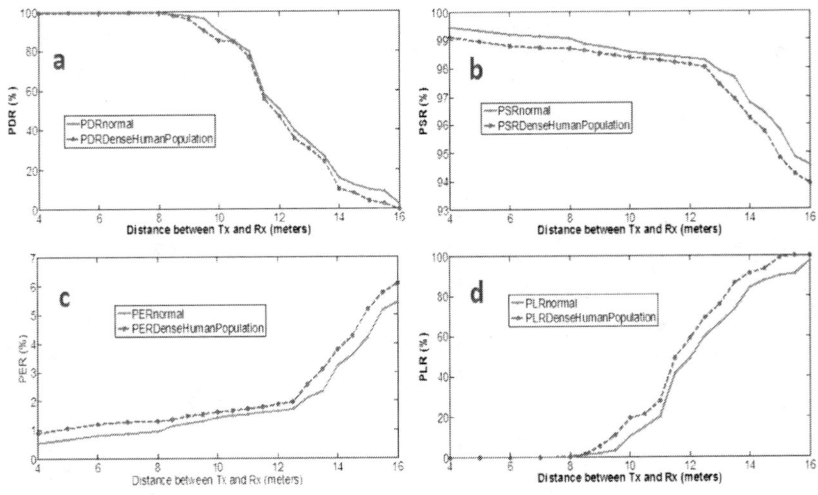

Figure 17. Attenuation loss caused by two different scenarios; one is regular, and other is in a densely populated smart building, measured between transmitter and receiver.

5. INTERFERENCE AND ATTENUATION ISSUES AND MITIGATION

Digi XBee ZigBee RF device is based on the CSMA-CA but even with that it is not able to mitigate the interference sufficiently. To investigate the characteristics of interference affecting smart building network, we have set up a controlled noise environment. The controlled noise of different noise levels was applied, and the effect at the receiver end was measured [46].

The center frequency of smart building network operation is 2425 MHz. In the smart building environment usually we find the microwave oven, Wi-Fi, Bluetooth and Ham Radio, which operate under ISM spectrum. As shown in Figure 18, to produce controlled noise, we used ROHDE & SCHWARZ 835.8011.58 with different types of antennas, and spectrum analyzer Anritsu MS2127A were used for interference measurement. We did not consider antenna orientation; we only considered the noise source range and frequency band.

Figure 18. Controlled noise setup to investigate the interference characteristics of the smart building environment; (**a**) External noise generated by planner antenna; (**b**) External noise generated by directional antenna; (**c**) External noise generated by isotropic antenna and (**d**) External noise generated by small planner antenna.

In obtaining Figure 19, the microwave signal of amplitude 60 dBm occupying the ISM band frequency was applied. Initially, this microwave signal was applied at 2400 MHz, and the corresponding SNR value for ZigBee network was measured, but as the microwave signal operating frequency was moving close to ZigBee network operating frequency, the SNR value was degrading. When it reached the 2425 ± 14 MHz, it had disturbed the whole signal, as we can see by Figure 19, and packet reception became zero. With a similar approach for Wi-Fi, Ham Radio, and Bluetooth, the disturbance regions were identified at 2425 ± 11 MHz and 2425 ± 12 MHz, respectively, whereas Bluetooth did not affect SNR significantly.

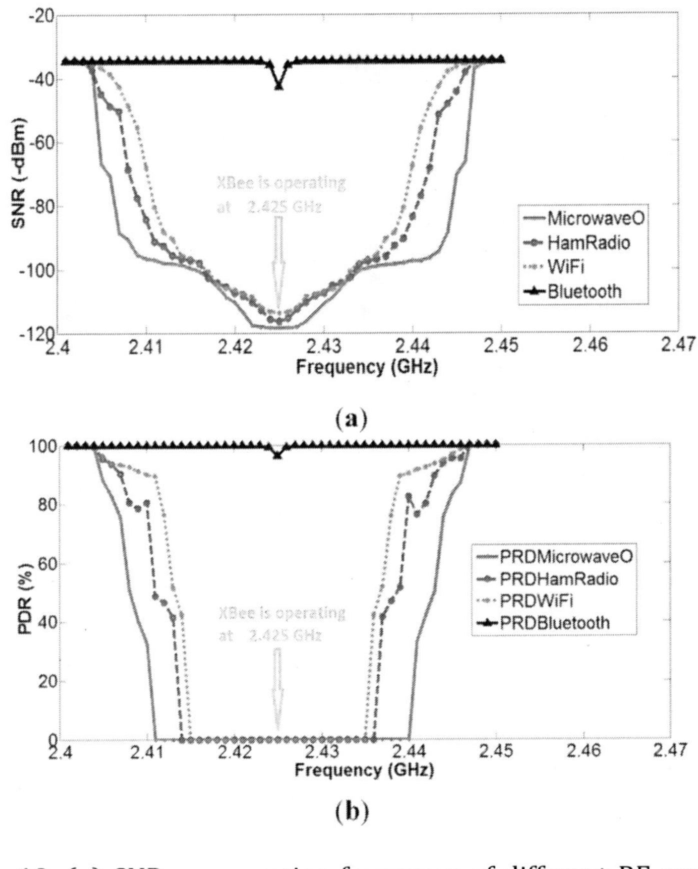

Figure 19. (**a**) SNR *vs.* operating frequency of different RF sources in ISM band; and (**b**) PDR *vs.* operating frequency of different RF sources in ISM band to investigate effect of interferences on ZigBee smart building network.

One of the best ways is to check the noise level in the ISM band and identify the frequency where the interference is least each time before smart system installation, and select that particular channel for smart building operation. The channel frequency selection is made possible by XBee boot settings.

Attenuation Mitigation

In the household environment, people usually place different types of material here and there regardless of sensor nodes. When the household stuff is placed near to sensor nodes, it becomes an obstruction for wireless communication. So, the placement of sensor nodes should be done in such a way where it gets least affected by household stuff, and the inhabitants should be made aware to avoid the placement of household stuff near sensor nodes. Changing the building material to reduce the attenuation is not an option for the researcher, but the appropriate placement of sensor nodes in the household environment can reduce the attenuation loss to a significant extent.

6. CONCLUSIONS

Wireless Sensor Networks and Internet of Things-based smart home is becoming an important ambient-assisted living environment for individuals, where necessary care can be provided at the time of need, and wellness can be measured and predicted. An old home built in 1938 had been converted into a smart home with the help of sensing technology and was in operation since May 2013. The expertise and knowledge of smart home have been explored to extend it to an intelligent building. ISM band interference and attenuation issues have been considered to observe the effectiveness of wireless communication and placement of wireless nodes. More results will be reported in our future works.

ACKNOWLEDGMENTS

The authors acknowledge the support and resources of the School of Engineering and Advanced Technology, Massey University, New Zealand.

AUTHOR CONTRIBUTIONS

NS has conceptualized the smart home monitoring system and wellness function generation. The approach of extension to building environment, loss mitigation and necessary experimental work followed by data analysis is done by HG. NS have provided important intellectual suggestions for this extension. SM and XG are supervisors and responsible for research inputs and editing the manuscript. HG is responsible for first draft of the paper. All authors contributed for editing and improvement of the paper. SM is the corresponding author.

CONFLICTS OF INTEREST

The authors declare no conflict of interest.

REFERENCES

1. Younis, M.; Senturk, I.F.; Akkaya, K.; Lee, S.; Senel, F. Topology management techniques for tolerating node failures in wireless sensor networks: A survey. *Comput. Netw.* **2014**, *58*, 254–283.

2. Challoo, R.; Oladeinde, A.; Yilmazer, N.; Ozcelik, S.; Challoo, L. An Overview and Assessment of Wireless Technologies and Co-existence of ZigBee, Bluetooth and Wi-Fi Devices. *Proced. Comput. Sci.* **2012**, *12*, 386–391.

3. Hwang, K.-I.; Choi, B.-J.; Kang, S.-H. Enhanced self-configuration scheme for a robust ZigBee-based home automation. *IEEE Trans. Consum. Electron.* **2010**, *56*, 583–590.

4. Byun, J.; Jeon, B.; Noh, J.; Kim, Y.; Park, S. An intelligent self-adjusting sensor for smart home services based on ZigBee Communications. *IEEE Trans. Consum. Electron.* **2012**, *58*, 794–802.

5. IEEE 802.15™: Wireless Personal Area Networks (PANs). Available online: http://standards.ieee.org/about/get/802/802.15.html (accessed on 1 January 2015).

6. The Internet of Things. http://share.cisco.com/internet-of-things.html (accessed on 7 January 2015).

7. IEEE Talks IOT. Available online: http://iot.ieee.org/articles-publications/ieee-talks-iot/88-ieee-talks-iot-oleg-logvinov.html (accessed on 7 January 2015).

8. Bell, C.; Fausset, C.; Farmer, S.; Nguyen, J.; Harley, L.; Fain, W.B. Examining social media use among older adults. In Proceedings of the 24th ACM Conference on Hypertext and Social Media, Paris, France, 1–3 May 2013; pp. 158–163.

9. Dawadi, P.N.; Cook, D.J.; Schmitter-Edgecombe, M.; Parsey, C. Automated assessment of cognitive health using smart home technologies. *J. Tech. Health Care* **2013**, *21*, 323–343.

10. Rantz, M.; Alexander, G.; Galambos, C.; Vogelsmeier, A.; Popejoy, L.; Flesner, M.; Lueckenotte, A.; Crecelius, C.; Zwygart-Stauffacher, M. Initiative to test a multidisciplinary model with advanced practice nurses to reduce avoidable hospitalizations among nursing facility residents. *J. Nurs. Care Qual.* **2013**, *29*, 1–8.

11. Place Lab MIT. Available online: http://web.media. mit.edu/~kll/A_PlaceLab_Sept1-2004.pdf (accessed on 10 January 2015).

12. Heinz, M.; Martin, P.; Margrett, J.A.; Yearns, M.; Franke, W.; Yang, H.-I.; Wong, J.; Chang, C.K. Perceptions of technology among older adults. *J. Gerontol. Nurs.* **2013**, *39*, 42–51.

13. Yang, H.-I.; Babbitt, R.; Wong, J.; Chang, C.K. A framework for service morphing and heterogeneous service discovery in smart environments. In *Impact Analysis of Solutions for Chronic Disease Prevention and Management*; Springer: Berlin/Heidelberg, Germany, 2012; pp. 9–17.

14. Chen, S.-Y.; Chang, S.-F. A review of Smart Living space development in a cloud computing network environment.*Comput. Aided Design Appl.* **2009**, *6*, 513–527.

15. Marculescu, D.; Marculescu, R.; Zamora, N.H.; Stanley-Marbell, P.; Khosla, P.K.; Park, S.; Jayaraman, S.; Jung, S.; Lauterbach, C.; Weber, W. Electronic textiles: A platform for pervasive computing. *IEEE Proc.* **2003**, *91*, 1995–2018. Available online: http://ieeexplore. ieee.org/xpl/login.jsp?tp=&arnumber=1246382&url=http%3A%2F

%2Fieeexplore.ieee.org%2Fxpls%2Fabs_all.jsp%3Farnumber%3D1246382 (accessed on 29 April 2015).

16. Introducing the BodyMedia SenseWear System. Available online: http://www.sensewear.com/solutions_bms.php(accessed on 10 January 2015).

17. Grossman, P. The LifeShirt: A multi-function ambulatory system monitoring health, disease, and medical intervention in the real world. *Stud Health Tech. Inf.* **2004**, *108*, 133–141.

18. Voirin, G. Working Garment Integrating Sensor Applications Developed within the PROeTEX Project for Firefighters Ubiquitous Computing in the Workplace. *Adv. Intell. Syst. Comput.* **2015**, *333*, 25–33.

19. Han, G.; Jiang, J.; Shu, L.; Niu, J.; Chao, H.-C. Management and applications of trust in Wireless Sensor Networks: A survey. *J. Comput. Syst. Sci.* **2014**, *80*, 602–617.

20. Ramesh, M.V.; Shanmughan, A.; Prabha, R. Context aware ad hoc network for mitigation of crowd disasters. *Ad Hoc Netw.* **2014**, *18*, 55–70.

21. Felisberto, F.; Fdez-Riverola, F.; Pereira, A. A Ubiquitous and Low-Cost Solution for Movement Monitoring and Accident Detection Based on Sensor Fusion. *Sensors* **2014**, *14*, 8961–8983.

22. Tunca, C.; Alemdar, H.; Ertan, H.; Incel, O.D.; Ersoy, C. Multimodal Wireless Sensor Network-Based Ambient Assisted Living in Real Homes with Multiple Residents. *Sensors* **2014**, *14*, 9692–9719.

23. Magaña-Espinoza, P.; Aquino-Santos, R.; Cárdenas-Benítez, N.; Aguilar-Velasco, J.; Buenrostro-Segura, C.; Edwards-Block, A.; Medina-Cass, A. WiSPH: A Wireless Sensor Network-Based Home Care Monitoring System. *Sensors* **2014**,*14*, 7096–7119.

24. Delahoz, Y.S.; Labrador, M.A. Survey on fall detection and fall prevention using wearable and external sensors.*Sensors* **2014**, *14*, 19806–19842.

25. Gusmeroli, S.; Piccione, S.; Rotondi, D. A capability-based security approach to manage access control in the Internet of Things. *Math. Comput. Model.* **2013**, *58*, 1189–1205.

26. Zhong, N.; Ma, J.H.; Huang, R.H.; Liu, J.M.; Yao, Y.Y.; Zhang, Y.X.; Chen, J.H. Research challenges and perspectives on Wisdom Web of Things. *J. Supercomput.* **2013**, *64*, 862–882.

27. Chen, Z.; Xia, F.; Huang, T.; Bu, F.; Wang, H. A localization method for the Internet of Things. *J. Supercomput.* **2013**, *63*, 657–674.

28. Sanchez, L.; Muñoz, L.; Galache, J.A.; Sotres, P.; Santana, J.R.; Gutierrez, V.; Ramdhany, R.; Gluhak, A.; Krco, S.; Theodoridis, E. SmartSantander: IoT experimentation over a smart city testbed. *Comput. Netw.* **2014**, *61*, 217–238.

29. Grieco, L.; Rizzo, A.; Colucci, S.; Sicari, S.; Piro, G.; di Paola, D.; Boggia, G. IoT-aided robotics applications: Technological implications, target domains and open issues. *Comput. Commun.* **2014**, *54*, 32–47.

30. Fabian, B.; Feldhaus, T. Privacy-preserving data infrastructure for smart home appliances based on the Octopus DHT. *Comput. Ind.* **2014**, *65*, 1147–1160.

31. Bravo, J.; Fuentes, L.; de Ipina, D.L. Theme issue: "Ubiquitous computing and ambient intelligence". *Pers. Ubiquitous Comput.* **2011**, *15*, 315–316.

32. Kim, W.H.; Lee, S.; Hwang, J. Real-time energy monitoring and controlling system based on Zigbee sensor networks. *Proced. Comput. Sci.* **2011**, *5*, 794–797.

33. Desai, A.S.; Konstam, M.A. Home Monitoring Heart Failure Care Does Not Improve Patient Outcomes Looking beyond Telephone-Based Disease Management. *Circulation* **2012**, *125*, 828–836.

34. Zhang, Y.; Huang, G.Q.; Qu, T.; Ho, O.; Sun, S. Agent-based smart objects management system for real-time ubiquitous manufacturing. *Robot. Comput. Integr. Manuf.* **2011**, *27*, 538–549.

35. Härmark, L.; van Grootheest, K. Web-based intensive monitoring: From passive to active drug surveillance. *Expert Opin. Drug Saf.* **2012**, *11*, 45–51.

36. RF Basics. http://community.arubanetworks.com/aruba/attachments/aruba/tkb@tkb/121/1/RF-Basics_Part1.pdf(accessed on 15 January 2015).

37. Suryadevara, N.K.; Mukhopadhyay, S.C.; Kelly, S.D.T.; Gill, S.P.S. WSN-Based Smart Sensors and Actuator for Power Management in Intelligent Buildings. *IEEE Trans. Mechatron.* **2014**.

38. Suryadevara, N.; Gaddam, A.; Rayudu, R.; Mukhopadhyay, S. Wireless sensors network based safe home to care elderly people: Behaviour detection. *Sens. Actuators A Phys.* **2012**, *186*, 277–283.

39. Suryadevara, N.K.; Mukhopadhyay, S.C. Sensor Activity Pattern (SAP) Matching Process and Outlier Detection. In*Smart Homes*; Springer: Berlin, Germany, 2015.

40. Suryadevara, N.K.; Mukhopadhyay, S.C. Wireless sensor network based home monitoring system for wellness determination of elderly. *IEEE Sens. J.* **2012**, *12*, 1965–1972.

41. Kelly, S.D.T.; Suryadevara, N.; Mukhopadhyay, S.C. Towards the implementation of IoT for environmental condition monitoring in homes. *IEEE Sens. J.* **2013**, *13*, 3846–3853.

42. Suryadevara, N.; Mukhopadhyay, S.C.; Wang, R.; Rayudu, R. Forecasting the behavior of an elderly using wireless sensors data in a smart home. *Eng. Appl. Artif. Intell.* **2013**, *26*, 2641–2652.

43. Dual Band Versatility. Available online: http://files.metageek.net/marketing/Wi-Spy_DBx/Wi-spy_DBx_medium.pdf(accessed on 17 January 2015).

44. Powerful RF Analyzer for Wi-Fi. Available online: http://www.streakwave.com/metageek/Chan_Pro_medium.pdf(accessed on 17 January 2015).

45. The Basics of Signal Attenuation. Available online: http://www.dataloggerinc.com/content/resources/white_papers/332/the_basics_of_signal_attenutaion (accessed on 17 January 2015).

46. XCTU. Available online: http://ftp1.digi.com/support/documentation/90001003_A.pdf (accessed on 18 January 2015).

2

QUALITY OF SERVICE IN WIRELESS SENSOR NETWORKS

Joseph E. Mbowe1, George S. Oreku2

1 The Nelson Mandela African Institution of Science and Technology, School of Computational Science and Communication Engineering, Arusha, Tanzania

2 Faculty of Economic Science and Information Technology, TIRDO/North West University, Vanderbijlpark, South Africa

ABSTRACT

The growing demand of usage of wireless sensors applications in different aspects makes the quality-of-service (QoS) to be one of paramount issues in wireless sensors applications. Quality of service guarantee in wireless sensor networks (WSNs) is difficult and more challenging due to the fact that the resources available of sensors and the various applications running over these networks have different constraints in their nature and re- quirements. Traditionally quality of service was focused on network level with concern in metrics such as delay, throughput, jitter e.c.t. In this paper we present appropriate metrics of QoS for WSN which involve service, re- liability and availability which ultimately facilitating in archiving qualitable service. We discuss the reverse look of QoS and hence present mathematically the three significant quality factors that should currently be taken into account in developing WSNs application quality services namely, availability, reliability and serviceability. We run experiments incorporating these three phenomenons (reliability, availability and serviceability—RAS) to demonstrate how to attain QoS which effectively improve reliability of the overall WSNs.

KEYWORDS

Wireless Sensor Network; QoS; Probability; Metrics; Serviceability

1. INTRODUCTION

Quality of service is an overused term with multiple meanings and perspectives from different research and technical communities [1]. QoS in WSNs can be viewed from two perspectives: application-specific and network. The former refers to QoS parameters specific to the application, such as sensor node measurement, deployment, and coverage and number of active sensor nodes. The latter refers to how the supporting communication network can meet application needs while efficiently using network resources such as bandwidth and power consumption.

With the recent technological developments of the wireless networks and multifunctional sensors with processing and communication capabilities, wireless sensor networks (WSNs) have been used in an increasing number of applications. WSNs can provide a more accurate or reliable monitoring service for different classes of applications [2,3]. Quality of service can be an important mechanism to guarantee that the distinct requirements for different classes of applications are met [4].

Traditional QoS mechanisms used in wired networks aren't adequate for WSNs because of constraints such as resource limitations and dynamic topology. One of the many challenges concerning wireless sensor networks (WSNs) is how to provide Quality of Service (QoS) parameter guarantees in real-time applications [5]. Therefore, middleware should provide new mechanisms to maintain QoS over an extended period and even adjust itself when the required QoS and the state of the application changes. Middleware should be designed based on trade-offs among performance metrics such as network capacity or throughput, data delivery delay, and energy consumption in order to provide QoS in Wireless Sensor Network.

1.1. QoS Concept

As defined in [6], Quality-of-Service is a set of service requirements to be met by the network while transporting a flow. "Here a flow is" a packet stream from source to a destination (unicast or multicast) with an associated Quality of Service (QoS) [6]. In other words, QoS is a measurable level of service delivered to network users, which can be characterized by packet loss probability, available bandwidth, end-to-end delay, etc. Such QoS can be provided by network service providers in terms of some agreement (Service Level Agreement, or SLA) between network users and service providers. For example, users can require that for some traffic flows, the network should choose a path with minimum 2M bandwidth.

1.2. QoS Metrics

For quality of service to be implemented, service requirements have to be expressed in some measurable QoS metrics. The well-known metrics include bandwidth, delay, jitter, cost, loss probability, etc. Different metrics may have different features. There are 3 types of metrics when talking about QoS: additive, multiplicative, and concave [7]. These can be defined as follows:

Let m (n_1, n_2) be a metric for link (n_1, n_2). For any path $P = (n_1, n_2, \cdots, n_i, n_j)$, metric m is: (Note here $n_1, n_2, n_3, \cdots, n_i, n_j$ represent network nodes)

- additive, if $m(P) = m(n_1, n_2) + m(n_2, n_3) + \cdots + m(n_i, n_j)$

Examples are delay, jitter, cost and hop-count. For instance, the delay of a path is the sum of the delay of every hop.

- multiplicative, if $m(P) = m(n_1, n_2) * m(n_2, n_3) * \cdots * m(n_i, n_j)$

Example is reliability, in which case $0 < m(n_i, n_j) < 1$.

- concave, if $m(P) = \min\{m(n_1, n_2), m(n_2, n_3), \cdots, m(n_i, n_j)\}$

Example is bandwidth, which means that the bandwidth of a path is determined by the link with the minimum available bandwidth.

2. QOS CHALLENGES IN SENSOR NETWORKS

Different from IP network, Sensor network naturally supports multiple service types, thus provides different QoS. The service types range from CBR (Constant Bit Rate) which guarantees bandwidth, delay and delay jitter, to UBR (Unspecified Bit Rate) which virtually provides no guarantees (just like today's "best-effort" IP network). While sensor networks inherit most of the QoS issues from the general wireless networks, their characteristics pose unique challenges. The following is an outline of design considerations for handling QoS traffic in wireless sensor networks.

Bandwidth limitation: A typical issue for general wireless networks is securing the bandwidth needed for achieving the required QoS. Bandwidth limitation is going to be a more pressing issue for wireless sensor networks. Traffic in sensor networks can be burst with a mixture of real-time and non-real-time traffic. Dedicating available bandwidth solely to QoS traffic will not be acceptable. A trade-off in image/video quality may be necessary to accommodate non-real-time traffic. In addition, simultaneously using multiple independent routes will be sometime needed to split the traffic and allow for meeting the QoS requirements. Setting up independent routes for the same flow can be very complex and challenging in sensor networks due energy constraints, limited computational resources and potential increase in collisions among the transmission of sensors.

Removal of redundancy: Sensor networks are characterized with high redundancy in the generated data. For unconstrained traffic, elimination of redundant data messages is somewhat easy since simple aggregation functions would suffice. However, conducting data aggregation for QoS traffic is much more complex. Comparison of images and video streams is not computationally trivial and can consume significant energy resources. A combination of system and sensor level rules would be necessary to make aggregation of QoS data computationally feasible. For example, data aggregation of imaging data can be selectively performed for traffic generated by sensors pointing to same direction since the images may be very similar. Another factor of consideration is the amount of QoS traffic at a particular moment. For low traffic it may be more efficient to cease data aggregation since the overhead would become dominant. Despite the complexity of data aggregation of imaging and video data, it can be very rewarding from a network performance point-of-view given the size of the data and the frequency of the transmission.

Energy and delay trade-off: Since the transmission power of radio is proportional to the distance squared or even higher order in noisy environments or in the nonflat terrain, the use of multi-hop routing is almost a standard in wireless sensor networks. Although the increase in the number of hops dramatically reduces the energy consumed for data collection, the accumulative packet delay magnifies. Since packet queuing delay dominates its propagation delay, the increase in the number of hops can, not only slow down packet delivery but also complicate the analysis and the handling of delay-constrained traffic. Therefore, it is expected that QoS routing of sensor data would have to sacrifice energy efficiency to meet delivery requirements. In addition, redundant routing of data may be unavoidable to cope with the typical high error rate in wireless communication, further complicating the trade-off between energy consumption and delay of packet delivery.

Buffer size limitation: Sensor nodes are usually constrained in processing and storage capabilities. Multi-hop routing relies on intermediate relaying nodes for storing incoming packets for forwarding to the next hop. While a small buffer size can conceivably suffice, buffering of multiple packets has some advantages in wireless sensor networks. First, the transition of the radio circuitry between transmission and reception modes consumes considerable energy and thus it is advantageous to receive many packets prior to forwarding them. In addition, data aggregation and fusion involves multiple packets. Multihop routing of QoS data would typically require long sessions and buffering of even larger data, especially when the delay jitter is of interest. The buffer size limitation will increase the delay variation that packets incur while traveling on different routes and even on the same route. Such an issue will complicate medium access scheduling and make it difficult to meet QoS requirements.

Support of multiple traffic types: Inclusion of heterogeneous set of sensors raises multiple technical issues related to data routing. For instance, some applications might require a diverse mixture of sensors for monitoring temperature, pressure and humidity of the surrounding environment, detecting motion via acoustic signatures and capturing the image or video tracking of moving objects. These special sensors are either deployed independently or the functionality can be included on the normal sensors to be used on demand. Reading generated from these sensors can be at different rates, subject to diverse quality of service constraints and following multiple data delivery models, as explained earlier. Therefore, such a heterogeneous environment makes data routing more challenging.

3. RELIABILITY, AVAILABILITY AND SERVICEABILITY

As Wireless Sensor Networks (WSNs) are expected to be adopted in many industrial, health care and military applications, their reliability, availability and serviceability (RAS) are becoming critical. In recent years, the diverse potential applications for wireless sensor networks (WSN) have been touted by researchers and the general press [8-10]. In many WSNs systems, to provide sufficient RAS can often be absorbed in the network cost. Nevertheless, as noticed early [11], network designers face "two fundamentally conflicting goals: to minimize the total cost of the network and to provide redundancy as a protection against major service interruptions."

For availability and serviceability, remote testing and diagnostics is needed to pinpoint and repair (or bypass) the failed components that might be physically unreachable. Severe limitations in the cost and the transmitted energy within WSNs negatively impact the reliability of the nodes and the integrity of transmitted data. The application itself will greatly influence how system resources (namely, energy and bandwidth) must be allocated between communication and computation requirements to achieve requisite system performance. The presentation below demonstrates how different application wireless sensor nodes can influence the resource usability:

Power states are states of particular devices; as such, they are generally not visible to the user. For example, some devices may be in the Off state even though the system as a whole is in the working state.

These states are defined very generically in this section to enable applications adopted in our approach. Many devices do not have all four power states defined. Devices may be capable of several different low power modes, but if there is no user-perceptible difference between the modes only the lowest power mode will be used. We define four power states according to advanced configuration power interface (ACPI) [12]:

Ready—(or busy) is when the system or device is fully powered up and ready for use.

Idle—is an intermediate system dependent state that attempts to conserve power. The CPU enters the idle state when no device activity has occurred within a machine defined time period. The machine won't return to busy state until a device raises a hardware interrupt or the machine accesses a controlled device.

Suspend—is the lowest level of power consumptions available in which memory preserves all data and operational parameters. The device won't perform any computations until it resumes normal activity, which it does when signal by an external event such as a button press, timer alarm, or receipt of request.

When Off—the device is powered down and inactive. Operational and data parameters might or might not be preserved in **Figure 1** shows the general current ranges for each operating state as well as the power distribution for a PDA class of devices. Cliff Brake affirms that the CPU accounts for approximately 30 percent of power and the screen 42 percent when backlit these percentages vary slightly with each PDA class [13]. In an idle state, the CPU loses nearly all current and the backlight is turned off, equating to an approximate 64 percent power reduction.

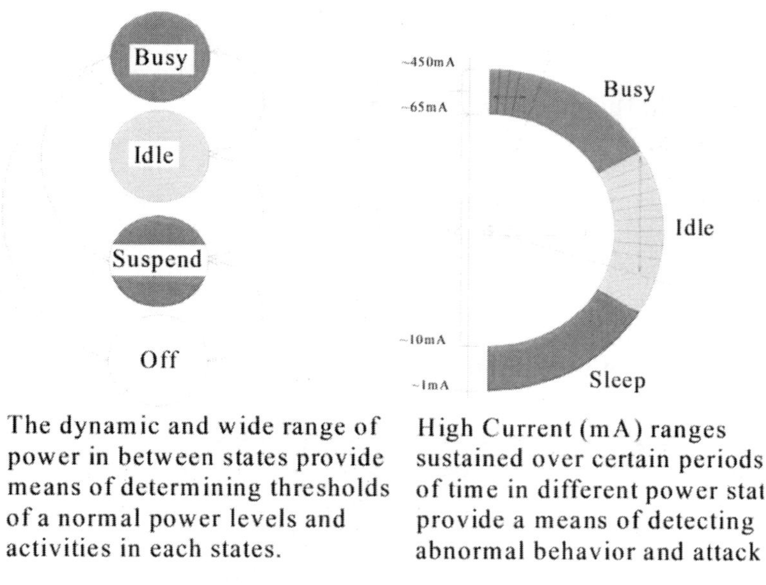

Figure 1. State power distribution (adapted from a Dell Axim) and battery-based intrusion detections (B-BID) power drain rate thresholds. The longer a threshold is held high in the busy and idle states, the greater the likehood that an anomalous activity is present.

This can be deceiving, however. In idle state, if the wireless local area network (LAN) card picks up a network request and transmits an acknowledgement, the CPU will consume power at a higher level. Worse yet, once on, the card might pick up multiple requests, and unless the user has altered the CPU's communication protocol, it will try to send multiple acknowledgements for each request. In addition, the power required to transmit is greater than it is to receive by approximately 1.5:1 [14,15].

The justification of idle state resource consumption can be only identified through worse or best scenarios as follow:

$$E = N\left(1 + r/t\right)^{T} \tag{1}$$

The inputs are the total number of nodes (N), threshold (r), the time taken (t), and total time (T). One of the purposes of a model such as this is to make predictions and try "What If?" scenarios. You can change the inputs and recalculate the model and you'll get a new answer. You might even want to plot a graph of the expected results (E) vs. time (T). In some cases, you may have a fixed results rate, but what do you do if the results rate is allowed to change? For this simple equation, you might only care to know a worst/best case scenario, where you calculate the expected value based upon the lowest and highest results rates that you might expect.

While examining WSN nodes and propose the necessary QoS required for increasing both the availability and serviceability of the system our approach is service oriented and was particularly motivated by recent proposals to define QoS (quality of service) for WSN. In one definition, QoS measures application reliability with a goal of energy efficiency [16]. An alternative definition equates QoS to spatial resolution [17]. This latter work also presented a QoS control strategy based on a Gur game paradigm in which base stations broadcast feedback to the network's sensors. QoS control is required for the assumption is that the number of sensors deployed exceeds the minimum needed to provide the requisite service.

This work presents two new techniques to maintain QoS under a variety of network constraints. We first adapt the proposed Gur game strategy to operate in energy poor environments then proposes a new, extremely low-energy control strategy based on individual feedback in a random access communication system. In particular, our work is applicable to networks that are deployed in remote, harsh environs (e.g., space applications). Such networks are constrained by (1) high die-off rates of nodes and (2) inability to be replenished. The performance of the proposed algorithms is demonstrated throughout using numerical examples as follows (2) and (3):

$$\text{Reliability} = 1 - \frac{t}{\text{Mean\_time\_between\_failure}} \tag{2}$$

Where m is a number of failed nodes within WSN.

n is number of nodes within WSN and $M\%$ is possible percentage of failed nodes within given WSN.

$$\text{Serviceability} = 1 - \exp \times -\left(\frac{t}{\text{Mean\_time\_to\_repair}} \right)$$

4. CALCULATING PROBABILITY OF NODES AVAILABILITY IN WSN

The availability of several implementations is derived from Equation (3) above for Mean Time between Failure (MTBF) and Mean Time to Repair (MTTR). Due to the power issue and the unpredictable wireless network characteristics, it is possible that applications running on the sensor nodes might fail. Thus, techniques to improve the availability of sensor nodes are necessary. Estimated MTBF in our sensor nodes is based on the individually calculated failure rates for each component and the circuit board. Next, for the redundant system versions, if the failure rates (λ) of each redundant element are the same, then the MTBF of the redundant system with n parallel independent elements (i) [18] are taken as:

$$Availability = \frac{\text{Mean\_time\_between\_failure}}{\text{Mean\_time\_between\_failure} + \text{Mean\_time\_to\_repair}}$$

$$M\% = \frac{m \times 100\%}{n} \tag{3}$$

$$\text{Mean\_time\_between\_failure} = \sum_{i=1}^{n} \frac{1}{i\lambda} \tag{4}$$

The MTTR can be estimated by the sum of two values, referred to as Mean Time to Detect (MTTD) the failures and the Time to Repair (TTR)

(MTTR = MTTD + TTR). Notice that this part might be severely affected by the network connections.

Considering the technique [19], where the consumer starts the reparation mechanism by activating the local functional test. Once it completes, the test result is sent back to the consumer for analysis. If a failure occurs, the consumer will send the repair message to the sensor node and initialize the backup component. Acknowledgement is sent back to the consumer once the reparation is completed. If the message latency from the consumer to the target node is d seconds and the test time is c seconds, then we calculate MTTR as Equation (5):

$$\text{Mean\_time\_to\_repair} : 4d + c \tag{5}$$

For the sensor node without the Test Interface Module [19], consumer sends the measured data request command to the suspected sensor node. In order to check the data integrity, same request command will also send to at least two other nearby sensor nodes. The consumer compares the three collected streams of data and pinpoints the failed node. Once the failure is confirmed, consumer will notify the surrounding sensor node to take over the applications of the failed node. Once the failure is confirmed, consumer will notify the surrounding sensor node to take over the applications of the failed node. Again if the message latency from the consumer to the target node is d seconds, then MTTR is:

$$\text{Mean\_time\_to\_repair} \sim 8d \tag{6}$$

To estimate realistic MTTR numbers, we use study [20], where for WSNs Thermostat application with 64 sensor nodes is simulated. Due to the power and protocol requirements, the average latency of related messages is 1522s. By applying this to our MTTR estimations, the test time c is much smaller and can be neglected.

Reliability of a system is defined as the probability of system survival Equation (7) in a period of time. Therefore, using Poisson probability [21] implemented for WSNs we have as well estimate probability of "failed" situation for whole WSN in given time interval, e.g. for one day (24 hours) to demonstrate the reliability of our presented approach.

$$\text{Probability}(r) = \frac{m^r \times e^{-m}}{r!} \tag{7}$$

Where Probability (r) is a probability of failure system working with "r" failed nodes within WSN for given time interval, $r \geq 0$, m is a average number of failed nodes within WSN and e = 2.718...

For example, in average there are 3 failed nodes in WSN fo 24 hours. Then we calculate Probabilities of failure system working as:

$$\text{Probability}\left(\text{"}r\text{" fails\_for\_24\_hours}\right) = \frac{3^r \times e^{-3}}{r!}$$

$$\text{Probability}\left(\text{0\_fail\_for\_24\_hours}\right)$$
$$= P(0) = \frac{3^0 \times e^{-3}}{0!} = 0.0498$$

$$\text{Probability}\left(\text{1\_fail\_for\_24\_hours}\right)$$
$$= P(1) = \frac{3^1 \times e^{-3}}{1!} = 0.1494$$

$$\text{Probability}\left(\text{4\_fails\_for\_24\_hours}\right)$$
$$= P(4) = \frac{3^4 \times e^{-3}}{4!} = 0.1680$$

From this example, we can see that with progressive increase of fail nodes quantity of a WSN, the risk of unstable work also increases.

5. EXPERIMENTS AND EVALUATION

The discussion in this section will be about achieving two primary factors of dependability in WSNs applications, namely availability and reliability. In the classical definition, a system is highly available if the fraction of its downtime is very small, either because failures are rare, or because it can restart very quickly after a failure [22].

The performance of the proposed approach is demonstrated throughout using numerical examples. Reliability of a system is defined as the probability of system survival in a period of time. Since it depends mainly on the operating conditions and operating time, the metrics of Mean Time between Failure (MTBF) is used. For time period of duration t, MTBF is related to the reliability as follows [19]:

$$\text{Mean\_time\_between\_failure} \tag{8}$$

Availability of a system is closely related to the reliability, since it is defined as the probability that the system is operating correctly at a given time. Dependence availability and reliability on MTBF presented on **Figure 2**. Calculating availability is related to MTBF and Mean Time to Repair (MTTR) by the following relation [19]:

$$\text{Availability} = \frac{\text{Mean\_time\_between\_failure}}{\text{Mean\_time\_between\_failure} + \text{Mean\_time\_to\_repair}} \tag{9}$$

Considering availability of each node in isolation, from Equation (9), the MTTR should be minimized, while MTBF should be maximized. While MTBF is given by manufacturing practices and components used, the value of MTTR can be controlled by both individual node and network design.

$$M\,\% = \frac{m \times 100\%}{n} \tag{10}$$

where m is a number of failed nodes within WSN, n is number of nodes within WSN and $M\%$ is possible percentage of failed nodes within given WSN.

Serviceability of a system is defined as the probability that a failed system will restore to the correct operation. Serviceability is closely related to the repair rate and the MTTR [19].

$$\text{Serviceability} = 1 - \exp \times \left(\frac{t}{\text{Mean\_time\_to\_repair}} \right) \tag{11}$$

A fundamental service in sensor networks is the determination of time and location of events in the real world. This task is complicated by various challenging characteristics of sensor networks, such as their large scale, high network dynamics, restricted resources, and restricted energy. We use Hawk sensor nodes for determination time of data transmitting in fulfilling the QoS under these constraints. We illustrate the practical feasibility to our approaches by concrete application of real sensor nodes (Hawk Sensor Nodes) to our experiments and the results of availability and reliability of sensor nodes to reveal QoS from our experiment can be seen on **Figure 2** above.

In any system one must consider the reliability of its components when ascertaining overall system performance. Thus our question was whether the proposed strategy performed adequately for various levels of sensor reliability. Equation (2), does not include any information regarding expected sensor life and thus assumes static network resources, which is clearly not the case in WSNs. For example, sensors may fail at regular intervals due to low reliability, due to cost driven design choices, environmentally caused effects (especially in harsh environments), loss of energy, etc.

We measured the processing throughput, i.e., the number of data transmitted events that each phase is able to process per second and time taken to transmit these data within selected sensor nodes, as can be seen in graph presentation in **Figure 3**.

We plot the node availability versus average latency, which lumps together the characteristics of the channel, the number of retransmission retries on the failure, as well as the node-dependent features such as retransmission timeouts in **Figure 4**.

In **Figure 5**, we examine WSNs nodes to transmit the data in evaluating (RAS). Two sensor nodes with 32 size byte were used for estimating connection time with different transmitting rate. With 0.0625 t/s, we were able to connect 32 packets. To ask one sensor node to transmit the data we need 2 data packets (one for asking, another one for receiving the answer). To estimate Time to connection we have to transmit only two packets. Number of packets = file size/packet size. Time = number of packets/ data transmitting rate. This can be used to propose the necessary infrastructure required for increasing both the availability and serviceability of the system, in spite of the absence of a reliable transport layer. Hence this can be used to analyze and detect delay, delivery, performance or energy consumptions.

6. CONCLUSIONS

One primordial issue in WSN is to satisfy application QoS requirements while providing a high-level abstraction that addresses good service. Notice that although we consider primarily testing in the laboratory, the proposed solutions can easily be applied to testing in factory with large size of Sensor network applications.

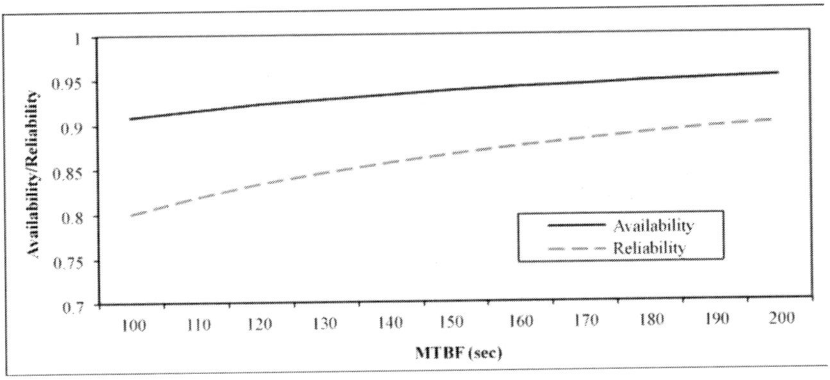

Figure 2. Dependence availability and reliability on MTBF.

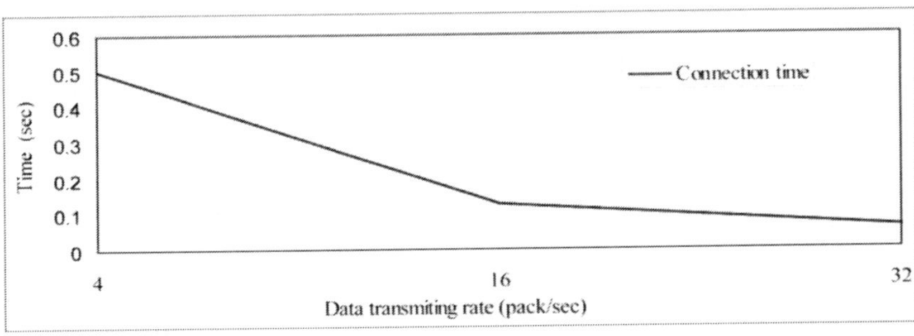

Figure 3. Connection time for 4/16/32 pack/sec.

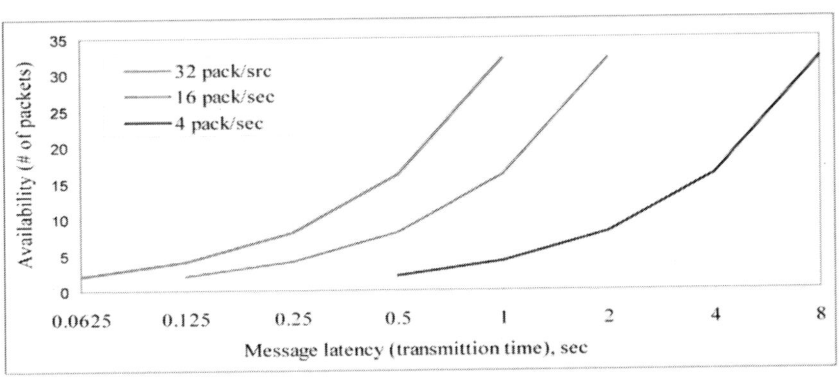

Figure 4. Availability of a Node in WSN.

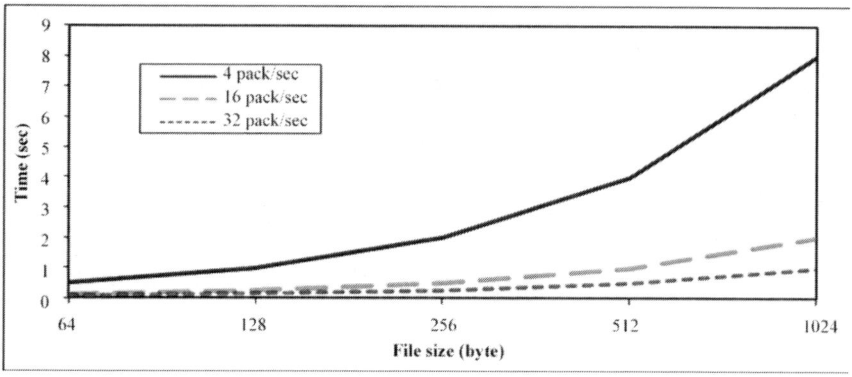

Figure 5. Transmitting time in different number of packets to access (RAS).

With the proposed approach, such tests can be easily parallelized by applying wireless broadcast to many nodes at once. As a result, the proposed approach can be used in variety of testing scenarios.

In this paper, QoS in WSN has been proposed through reliability availability and serviceability metrics, Using mentioned components we have evaluated QoS and system-level test using sensor nodes.

However our finding found that effects of traditional metrics (delay, throughput, jitter e.c.t.) place a lot of burden on the QoS of the overall system thus decreasing performance.

REFERENCES

1. D. Chen and P. K. Varshney, "QoS Support in Wireless Sensor Networks: A Survey," Proceedings of the International Conference on Wireless Networks (ICWN 04), Las Vegas, 21-24 June 2004, pp. 227-233.

2. J. Yick, B. Mukherjee and D. Ghosal, "Wireless Sensor Network Survey," Computer Networks, Vol. 52, No. 12, 2008, pp. 2292-2330.

3. J. L. Lu, W. Shu and W. Wu, "A Survey on Multipacket Reception for Wireless Random Access Networks," Journal of Computer Networks and Communications, Vol. 2012, 2012, 14 p.

4. D. Chen and P. K. Varshney, "QoS Support in Wireless Sensor Networks: A Survey," Proceedings of the International Conference on Wireless Networks, Las Vegas, 2004, pp. 609-619.

5. M. Z. Hasan and T. C. Wan, "Optimized Quality of Service for Real-Time Wireless Sensor Networks Using a Partitioning Multipath Routing Approach," Journal of Computer Networks and Communications, Vol. 2013, 2013, 18 p.

6. E. Crawley, R. Nair, B. Rajagopalan and H. Sandick, "A Framework for QoS-Based Routing in the Internet," RFC 2386, Internet Eng. Task Force, 1997.

7. S. G. Chen, "Routing Support for Providing Guaranteed End-to-End Quality-of-Service," Ph.D. Thesis, University of Illinois at Urbana-Champaign (UIUC), Champaign, 1999.

8. G. Pottie and W. Kaiser, "Wireless Integrated Network Sensors," Communications of the ACM, Vol. 43, No. 5, 2000, pp. 51-58.

9. D. Estrin, L. Girod, G. Pottie and M. Srivastava, "Instrumenting the World with Wireless Sensor Networks," Proceedings of International Conference Acoustics, Speech and Signal Processing (ICASSP 2001), May 2001, pp. 2675-2678.

10. MIT Technology Review, "10 Emerging Technologies That Will Change the World," MIT's Magazine of Innovation Technology Review, 2003. www.technologyreview.com

11. R. F. Rey, "Engineering and Operations in the Bell System," Bell Labs, Murray Hill, 1977.

12. X. Wang, W. Gu, K. Schosek, S. Chellappan and D. Xuan, "Sensor Network Configuration under Physical Attacks," Technical Report Technical Report (OSU-CISRC-7/04- TR45), The Ohio-State University, Columbus, 2004.

13. X. Wang, W. Gu, S. Chellappan, D. Xuan and T. H. Laii, "Search-Based Physical Attacks in Sensor Networks: Modeling and Defense," Technical Report, The OhioState University, Columbus, 2005.

14. H. Chan, A. Perrig and D. Song, "Random Key Predistribution Schemes for Sensor Networks," Proceedings of the 2003 IEEE Symposium on Security and Privacy, Washington DC, 2003, p. 197.

15. Y. Desmedt and S. Jajodia, "Redistributing Secret Shares to New Access Structures and Its Applications," Technical Report ISSE TR-97-01, George Mason University, Fairfax, 1997.

16. M. Perillo and W. Heinzelman, "Providing Application QoS through Intelligent Sensor Management," 1st Sensor Network Protocols and Applications Workshop (SNPA 2003), Anchorage, 11 May 2003, pp. 93-101.

17. R. Iyer and L. Kleinrock, "QoS Control for Sensor Networks," IEEE International Communications Conference (ICC 2003), Anchorage, 11-15 May 2003, pp. 517-521.

18. Arora, et al., "Extreme Scale Wireless Sensor Networking," Technical Report, 2004.http://www.cse.ohio-state.edu/exscal/

19. M. W. Chiang, Z. Zilic, K. Radecka and J.-S. Chenard, "Architectures of Increased Availability Wireless Sensor Network Nodes," ITC International Test Conference, Vol. 43, No. 2, 2004, pp. 1232-1241.

20. Headquarters, Department of the Army, "Reliability/Availability of Electrical & Mechanical Systems for Command, Control, Communications, Computer, Intelligence, Surveillance, and Reconnaissance Facilities," Department of the Army, Washington DC, 2003.

21. M. Eddous and R. Stansfield, "Methods of Decision Making," UNITY, Audit, 1997.

22. J. C. Knight, "An Introduction to Computing System Dependability," Proceedings of the 26th International Conference on Software Engineering (ICSE'04), Scotland, 2004, pp. 730-731.

3

ADAPTIVE CONTROL OF THE PACKET TRANSMISSION PERIOD WITH SOLAR ENERGY HARVESTING PREDICTION IN WIRELESS SENSOR NETWORKS

Kideok Kwon, Jihoon Yang and Younghwan Yoo *

Department of Computer Engineering, Pusan National University, Busandaehak-ro 63beon-gil Geumjeong-gu, Busan 609-735, Korea

ABSTRACT

A number of research works has studied packet scheduling policies in energy scavenging wireless sensor networks, based on the predicted amount of harvested energy. Most of them aim to achieve energy neutrality, which means that an embedded system can operate perpetually while meeting application requirements. Unlike other renewable energy sources, solar energy has the feature of distinct periodicity in the amount of harvested energy over a day. Using this feature, this paper proposes a packet transmission control policy that can enhance the network performance while keeping sensor nodes alive. Furthermore, this paper suggests a novel solar energy prediction method that exploits the relation between cloudiness and solar radiation. The experimental results and analyses show that the proposed packet transmission policy outperforms others in terms of the deadline miss rate and data throughput. Furthermore, the proposed solar energy prediction method can predict more accurately than others by 6.92%.

KEYWORDS

WSN; energy harvesting; solar energy; packet transmission period control

1. INTRODUCTION

Energy management is one of the essential factors in an embedded system due to the energy limitation. In wireless sensor networks (WSNs), sensor nodes collect various environmental data, such as temperature, humidity, sound and pressure. The faster the sensed data are sent to the server, the more valuable the data are, as they reflect the current situation more accurately. However, the data may not be delivered always the minute that it is collected due to the energy constraint of sensor nodes. A WSN suffers from the problem of small device dimensions. The device dimensions are related to the energy management, since a large battery cannot be equipped. Although many researchers have studied how to improve energy efficiency and how to enhance battery capacity, the lifetime of sensor nodes is still limited. Thus, an alternative technology is garnering attention, which is an energy harvesting system that acquires energy from various environmental sources, like the Sun, wind or pressure [1]. The past energy management in WSN focused on saving energy to prevent battery discharge. On the other hand, in the energy harvesting system, it is important to keep an energy-neutral policy, which means that the sensor node permanently maintains the performance demanded by user applications without power dissipation [2].

For an effective energy neutral policy, the system should be able to accurately predict the amount of energy that will be harvested in the future. The prediction methods are categorized depending on the type of utilized information. Some research predicts the current energy harvesting efficiency based on the harvested energy in the past. The others use weather information. These kinds of prediction methods provide better performance than the former, since they consider weather factors that are changing quickly in a short time. This prediction, however, requires high computational complexity and needs to sense environmental data for a long time to find a particular coefficient, although it is accurate within a variety of times ranging from three hours to a week.

This paper consists of two parts: energy harvesting prediction and data transmission period control. First, we predict the amount of harvested energy for the next three hours by using simple weather information. While existing prediction methods require a lot of weather factors, our method leverages just a few weather factors, which are the information of the past insolation and the cloud amount. From the relation between the two kinds of information, the amount of future solar radiation is deduced, so the proposed prediction method need not have a high computational or sensing overhead despite its high accuracy.

Second, we propose ACSE (adaptive control of packet transmission period with solar energy harvesting prediction) to control the packet transmission period based on our energy harvesting prediction. The basic philosophy is very simple: when energy is harvested plentifully, sensor nodes increase the throughput by shortening the transmission period. In the opposite case, they reduce energy consumption by lengthening the transmission period. ACSE aims to maintain the performance permanently while enhancing network throughput in the case of abundant solar insolation. It is easier to estimate solar energy harvesting than other renewable energy, because it has obvious features, like the distinct periodicity over a day. Such a feature is used by sensors to get a balance between throughput and energy consumption. As a result, the sensor node can maximally utilize the battery, relieving the worry about their battery dissipation.

The rest of the paper is organized as follows: Section 2 examines the related work about energy harvesting prediction and the packet scheduling policy considering the amount of harvested energy. Section 3 proposes a solar energy prediction method based on cloudiness and analyzes its accuracy. Section 4 suggests our transmission period control policy based on solar energy prediction. In Section 5, the simulation result shows that our method has better energy efficiency and meets the deadline of real-time works relatively well. Finally, Section 6 concludes this paper.

2. RELATED WORK

2.1. Solar Energy Harvesting Prediction Model

EWMA (energy prediction model using exponentially weighted moving-average) uses the profile about the amount of harvested energy in the past. The basic idea is that the amount of harvested energy does not change remarkably in a short time. EWMA splits a day into 48 sections from zero to 47, and each section is 30 min long. When x_k denotes the amount of the energy harvested at the k-th section and a is a weighting factor, the average of the harvested energy at each section is as in Equation (1). $\overline{x_k}$ means the k-th variance of the estimated harvesting energy. Here, how to decide the weighting factor a is very important, because the prediction accuracy totally depends on it. However, EWMA cannot dynamically react to environmental change.

$$\overline{x_k} = a\overline{x_{k-1}} + (1-a)x_k \tag{1}$$

The method called cloudy computing [3] predicts the amount of harvested solar energy, using weather information. It is shown that EWMA is not accurate enough for the recent future within 3 h. This method uses two main coefficients to predict harvested energy. One coefficient considers the location of the specific region, since the solar insolation is affected by the geographical location on the Earth. The other indicates how clear the sky is. It is insisted that the amount of harvested energy may be linearly proportional to the clearness. However, we will prove that this is not so simple through the thorough investigation of the past data.

Most WSN research focuses on reducing the energy consumption of sensors, but does not consider the future harvested energy. For instance, SEA-OR (Spectrum and Energy Aware Opportunistic Routing) [4] considers the physical location, the residual energy level and the link reliability to decide the back off-time of the sensor nodes, but future energy was not been taken into account. On the other hand, our method can effectively manage the energy in sensor batteries by considering the energy harvested in the near future, preventing the battery dissipation and harvested energy dumping due to the battery being fully charged.

2.2. WSN Policy Based on Energy Harvesting Prediction

Hsu *et al.* [5] suggests a power control policy in energy harvesting systems. For energy neutrality, they control the duty cycle of sensor nodes by making them switch between the active and the sleep mode repeatedly. However, they consider only constant data generation, so this is not suitable for real-time services, such as intrusion detection and volcano eruption monitoring.

LSA (lazy scheduling algorithm) [6] schedules packet transmission considering not only energy, but also the packet deadline. The authors insist on using lazy scheduling to solve the problem of the lack of energy, which can occur if greedy scheduling is applied to energy harvesting networks. However, the method may often miss the deadline of burst data.

The papers above have in common the assumption of constant data generation only. This assumption does not match real-time applications in WSN. Furthermore, they assume that the amount of the harvested energy can be predicted very accurately. However, this is dangerous, since solar energy significantly fluctuates depending on the weather. Our study relieves those assumptions and overcomes such limitations.

3. ENERGY HARVESTING PREDICTION

The proposed scheme predicts the future insolation by analyzing the relation between the past insolation and the cloud amount. Cloudy computing [3] has to collect the sunshine intensity at a specific area for several months to predict harvested energy. On the contrary, our method brings past weather information from KMA (Korea Meteorological Administration), as shown in Figure 1, without collecting this large amount of data. Thus, this system can be quickly deployed and used in any region without much advance preparation. Furthermore, the server estimates the future insolation based on the KMA data on behalf of sensor nodes and broadcasts it to the sensors. Therefore, each sensor does not have to receive a large amount of data from KMA and can save its energy for computation. In addition, since the server was designed to deal with weather data presented in the XML format (KMA provides the data in XML format), our system can be used in any country if their meteorological administration provides the weather data in an XML-compatible format. However, there may be an issue regarding the accuracy of the KMA data, since KMA provides data, not for the small region covered by a WSN, but for a large area, like a city. Moreover, the KMA data are sensed every hour on the hour, not at every instant. Currently, we argue that this kind of inaccuracy is a trade-off for the simple architecture and easy usage of the proposed system. Later, if the data for the difference between KMA information and sensed values are accumulated for a long time, the inaccuracy can be overcome by adopting a linear regression method.

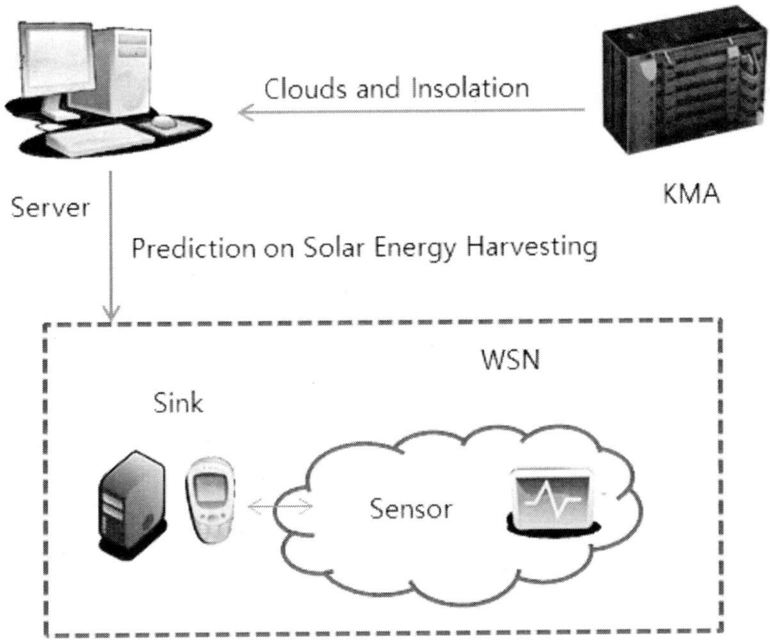

Figure 1. The system overview.

KMA measures the cloud amount and insolation every hour. Furthermore, they provide their own prediction on cloud amount and insolation for the next three hours. The amount of cloud is scaled from zero to 10. The closer to 10 it is, the more clouds there are. Figure 2 shows the distribution of pairs of cloudiness and insolation in June from 2003 to 2012. We can see that the insolation is tightly related to the amount of cloud. The more cloud there is, the lower the insolation intensity is. Our method predicts the insolation intensity by using this relationship. This procedure consists of four steps. Let us examine each step with an example:

- Decide the date and time we want to predict the solar insolation.

The time range for which we can predict insolation is limited to three hours in the future, because we can acquire only three-hour prediction on cloudiness from KMA. Let us assume that we want to predict the insolation in Busan, Korea, at 3 p.m., 2 August 2013.

- Get the past data on insolation and cloudiness from KMA.

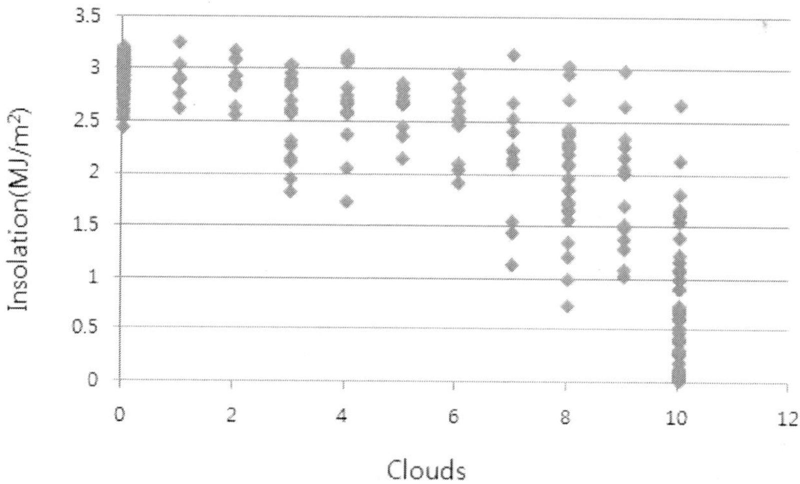

Figure 2. The insolation depending on the cloud amount in June from 2003 to 2012.

The insolation intensity is largely affected by the Sun's altitude, and the altitude changes with the Earth's revolution around the Sun. Thus, we get the data of 20 days around the targeting date or, in the previous example, the data of 10 days before and after 2 August (from 23 July to 12 August) are used to predict the future insolation. We also need to decide how many years' data will be used; then, we found out, through our experiment, that the prediction accuracy is not improved linearly in proportion to the amount of data. Therefore, we decide to get the data of the recent 10 years, as the prediction accuracy is not noticeably improved although more than 10 years' data are used.

- Compute the average of insolation over the days that were in a similar situation in the past.

As mentioned above, we got a total of 200 data (10 years × 20 days/year) from the KMA database, but we do not use all of the data. First, we filter out the days for which the cloudiness situation was different from the target instant. If the degree of cloudiness was predicted as seven by KMA at the aforementioned instant, 3 p.m., 2 August 2013, we extract only the data of the days whose cloudiness was seven. Figure 3 shows this screening result, with the number of days being counted for each quantized insolation intensity value (MJ/m^2). For instance, the number of days for which the insolation value was between 1.8 and 2.1 is five. Eventually, we can estimate the insolation at the target instant by averaging all of the insolation intensity values in Figure 3.

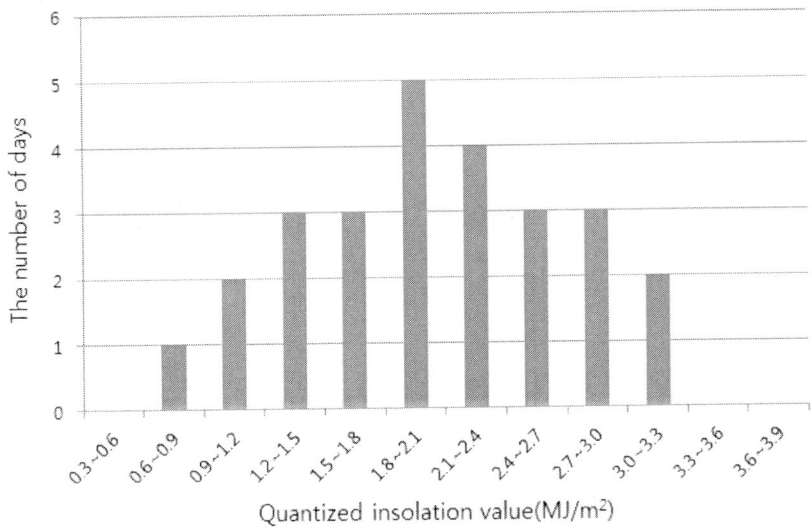

Figure 3. Quantized insolation distribution.

- Complement the estimated insolation according to the required confidence level.

In the last step, the average of the previous insolation intensities was adopted as the predicted insolation at the target instant. However, it is a little dangerous to use this value, as it is in the real-time embedded system, because the real insolation can be weaker than the estimated value. In this case, the system consumes more energy than the harvested one, resulting in the death of the real-time system. Therefore, we need to prevent the dissipation of the entire amount of energy by adopting a small conservative value as the estimated insolation. We calculated the confidence intervals of the estimated insolation with various confidence levels, $\alpha = 0.7$, 0.8 and 0.9, as follows:

$$X \sim N(\mu, \sigma^2),\ f(x) = \frac{1}{\sigma\sqrt{2\pi}} e^{-(x-\mu)^2/\sigma^2} \tag{2}$$

$$max\ a, P(X \geq a) \geq \alpha \tag{3}$$

Figure 4 compares the actual insolation with the original estimated value and the lower bound of each confidence interval. The actual values are always greater than the lower bounds with the confidence level = 0.9, while they go beyond the confidence intervals associated with the other confidence levels sometimes. As a result, it will be safe to adopt the lower bound of the confidence level = 0.9 as the estimated value. However, note that too conservative policies yield low network throughput by forcing the system to save too much energy even when enough residual energy remains in the battery.

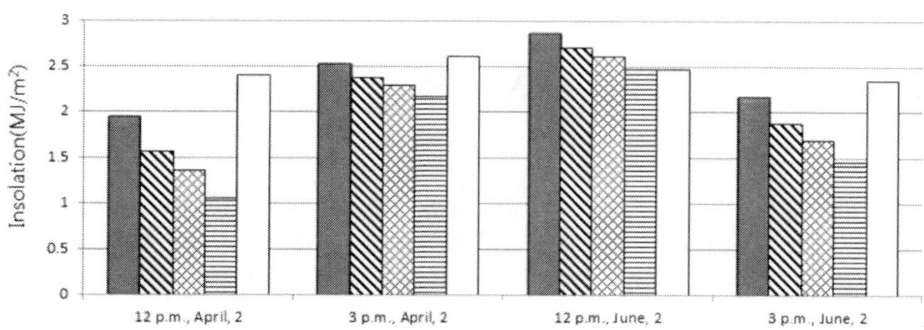

Figure 4. The comparison with the lower bounds of confidence intervals.

Figure 5 illustrates the insolation estimated by several schemes. For the proposed method, the original estimation values without the complementation using any confidence level are used. Although our method overestimates the insolation sometimes, the accuracy is very high in comparison with others. The accuracy of the proposed method and cloudy computing are 85.02% and 78.09%, respectively, the difference being 6.92%. Particularly, EWMA excessively underestimates the insolation, with only an accuracy of 40.52%, which may degrade the performance of the WSN too much.

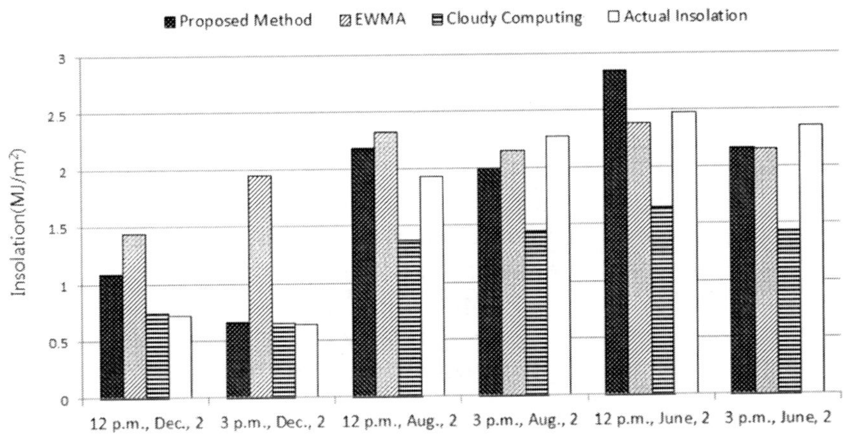

Figure 5. The comparison with existing algorithms.

4. ACSE: PACKET TRANSMISSION PERIOD CONTROL

This section suggests how to control the packet transmission period based on the prediction result in Section 3. Note that our transmission period control method, called ACSE, does not have to be combined with our own energy harvesting prediction all of the time. Provided that another prediction method is more accurate than our method, ACSE may give better performance by adopting it.

4.1. WSN and Task Model

Our model is similar to [7] in which a base station exists, and each sensor node has a periodic and an aperiodic task.Figure 6 shows a system consisting of a sensor, a transceiver, an energy harvesting module, a battery, a processor and memory. Each node sends packets to the BS (base station), and then, the BS forwards them to the server after data aggregation. The energy level in the energy storage module is between B^{min} and B^{max}. $H(t, h)$ in Equation (4) denotes the estimated energy to be harvested for h hours from t, and $p(t)$ is a function to predict the amount of harvested energy for one hour starting from time t.

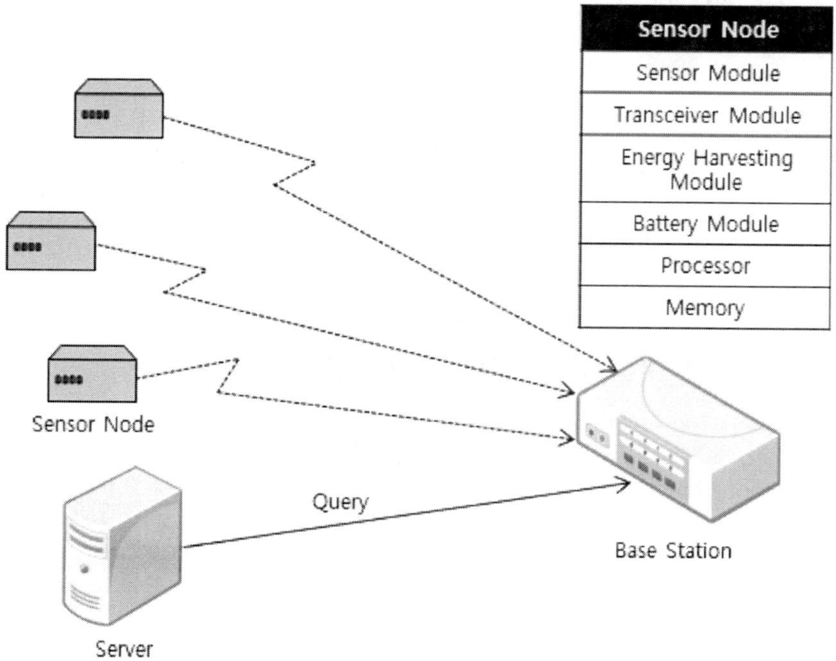

Figure 6. Energy harvesting sensor network model.

$$H(t,h) = \int_{t}^{t+h} p(x)dx,\ p(t) : a\ prediction function \\ = \sum_{x=t+1}^{t+1+h} p(x) \quad (4)$$

- An aperiodic task: A sensor node generates packets either when the condition is met or when it receives a server query. An example of the aperiodic task is: "If the temperature is over 30 °C, report its temperature immediately". When *E*and *R* denote the packet transmission energy and the average number of aperiodic events per hour, the total energy for*h* hours is defined as:

$$C(h) = E \times R \times h \quad (5)$$

- A periodic task: This task is to send a packet to the BS periodically. An example is: "Report the temperature every 30 min for the next 5 h". Each periodic task *x* is numbered from one to *n*. T_x denotes the transmission period of the periodic task x. Each task has a transmission deadline; thus, it has the minimum and the maximum period, T_{minx} and T_{maxx}. For the periodic task, the total energy consumption for h hours is computed as:

$$D(h) = hE \times \sum_{x} \frac{1}{T_x}, \quad T_x^{min} \leq T_x \leq T_x^{max} \tag{6}$$

4.2. Adaptive Transmission Period Control

The insolation intensity for a day typically has the form of a bell curve, like Figure 7, unless the weather changes abruptly during the daytime. The actual harvested energy is proportional to both the intensity and the conversion efficiency of the solar module. In the figure, H^{th} and c_e denote the energy that the sensor application requires and the convergence efficiency of the solar module.

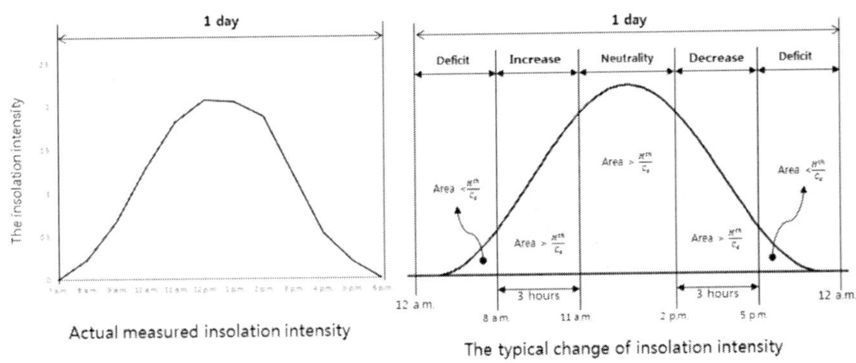

Figure 7. The insolation intensity for a day.

We divided the energy harvesting situations into four types: increase, decrease, neutrality and deficit. In the first three types, more energy is harvested than that required by the applications. They were subdivided into three types depending on the estimated change of insolation for the next three hours. If the insolation were to likely increase continuously, it is the increase type. The opposite case is the decrease type. If the insolation were guessed to fluctuate, it is the neutrality type. On the other hand, in the deficit type, more energy is used than the harvested energy.

Figure 8 depicts in detail how the energy harvesting type can be determined for the cases where the harvested energy is larger than the needed energy. This case includes the increase, decrease and neutrality types. For this decision, the gradient of the insolation change is calculated like this:

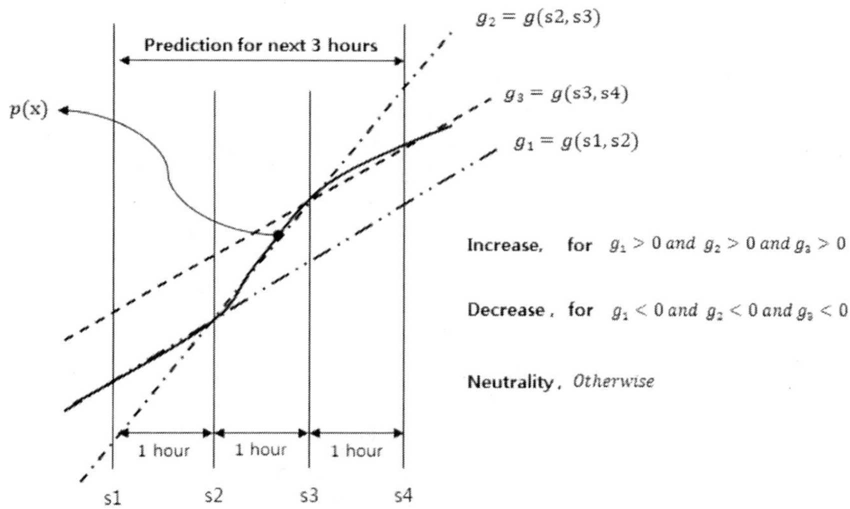

Figure 8. Decision for three energy harvesting types: increase, decrease and neutrality.

$$g(x_1, x_2) = \frac{p(x_2) - p(x_1)}{x_2 - x_1} \tag{7}$$

where $p(x_k)$ is the aforementioned prediction function on the insolation at x_k. The insolation is predicted every hour.

- Increase type: The amount of harvested energy is expected to continuously increase for the next three hours. Generally, it happens in the morning.

- Decrease type: Oppositely, as observed in the afternoon, the harvested energy is expected to continuously decrease for the next three hours

- Neutrality type: The harvested energy dose not continuously increase nor decrease for the next three hours. Namely, the harvested energy is predicted to fluctuate for the next three hours.

As mentioned earlier, if less energy is harvested than the energy needed, it is classified as the deficit type.

Our method, ACSE, dynamically controls the packet transmission of periodic tasks, depending on the energy harvesting types. The goal is to keep a WSN alive even when solar energy is not harvested at all, so that important aperiodic packets may be delivered in real time. At the same time, the throughput of periodic tasks should be maximized. Thus, ACSE adopts a different scheduling policy depending on the energy harvesting type. The scheduling policies will be described in detail later. Before that, the notations, listed in Table 1, are introduced.

Table 1. Notation definitions.

Symbol	Definition
α	The proportional factor for the energy ratio
$A(h)$	The total amount of harvested energy for the next h hours
$B^p(h)$	The energy that the periodic tasks can use for the next h hours
$C^{max}(h)$	The maximum energy that the aperiodic tasks may use for the next h hours
$D(h)$	The energy that the periodic tasks will use if they keep the current periods for the next h hours
h^r	The remaining time until the solar energy harvesting type is expected to get out of the deficit type

$$A(h) = H(t^{cur}, h) \quad (8)$$

$$B^p(h) = B^{cur} - B^{min} + A(h) - C^{max}(h) \quad (9)$$

The energy increasing and decreasing ratio can be figured out by using the equations in Equations (8)~(10), where $A(h)$ and $B^p(h)$ denote the total harvested energy and the maximum energy that the periodic tasks can use for the next h hours.B^{cur} is the current residual energy, and B^{min} is the minimum energy for a sensor node to keep running without harvested energy for some duration. In our method, this B^{min} was defined as the energy that the periodic tasks can use for one month with the longest possible period of packet generation, because the rainy season in Korea is rarely longer than one month. Thus, the minimum energy guarantees that a sensor node can provide its minimum service even if it cannot harvest energy at all for one month. Then, $C^{max}(h)$ denotes the maximum energy that aperiodic tasks may use for the next h hours. Actually, it is impossible to know how many aperiodic tasks will happen in advance, so we assume that they follow the Poisson distribution. Given an arrival rate λ, we compute the maximum number of events that can happen for the next h hours with a higher chance above 95%. Our method will be evaluated in the next section with various arrival rates.

$B^p(h)$ is the energy that the periodic tasks can use for the next h hours. It is compared with $D(h)$, which is the energy that the periodic tasks will use for the next h hours assuming that they maintain the current period. If $B^p(h)$ is less than $D(h)$, then a sensor node needs to use the longer period of its periodic task to prevent entire energy dissipation. Otherwise, the period can be shortened to increase the throughput, because the battery energy is enough. The ratio between $B^p(h)$ and $D(h)$ is defined as the proportional factor α in Equation (10). The energy consumption for the next h hours is increased or decreased depending on the value of α. If α is greater than one, the period of periodic tasks is shortened, because more available energy is expected for the next h hours.

$$\alpha = \frac{B^p(h)}{D(h)} \tag{10}$$

ACSE has four period control policies corresponding to the four different solar energy harvesting types.

- Neutrality policy: This policy is adopted in the neutrality type. Although the insolation intensity is not stable, the harvested energy is expected to be larger than or equal to the requested energy. Thus, the current periods are maintained for the next h hours.

- Increase policy: More and more energy is expected to be harvested continuously for the next h hours. Thus, each periodic task can transmit α times as many packets as the current number of packets. In Equation (11), T_x and $\overline{T_x}$ denote the current period and the new period, and n is the number of periodic task applications in the sensor node. The α is greater than one.

$$\overline{T_x} = \frac{1}{\alpha} \times T_x, T_x^{min} \leq \overline{T_x} \leq T_x^{max}, for x = 1, \cdots, n \quad (11)$$

- Decrease policy: Although the harvested energy is still greater than the needed energy, the insolation intensity is continuously weakened. Thus, the battery can be totally dissipated if the current period is kept for a long time. The periodic tasks lengthen their periods by using α smaller than one in Equation (11).

- Deficit policy: In the deficit type, less energy is harvested than the required energy, so staying alive until the deficit type ends is the goal. Instead of predicting for the next h hours, simply, ACSE firstly estimates h^r, which is the remaining time until the solar energy harvesting type will be changed into other types than the deficit type, by leveraging the weather information. Then, the proportional factor α that is computed between $B^p(h^r)$ and $D(h^r)$ is used to control the packet transmission period in Equation (11). Note that $B^p(h^r)$ and $D(h^r)$ denote the total energy that the periodic tasks can use and the energy required to keep the current period until enough energy can be harvested again.

$$B^p(h^r) = B^{cur} - B^{min} + A(h^r) - C^{max}(h^r)$$
$$\alpha = \frac{B^p(h^r)}{D(h^r)} \quad (12)$$

5. PERFORMANCE EVALUATION

Using MATLAB, we performed the simulation to evaluate ACSE in terms of energy efficiency and throughput. Therefore, before that, we needed to estimate the amount of energy that might be harvested during the simulation time. Setting the simulation to continue for five days from 8 a.m., 4 May 2014, we compared our estimated insolation intensity with the KMA's report. In short, our scheme in Section 3 could predict the future insolation fairly well with a difference of only 6.78% from the actual value. Each sensor node had three periodic tasks and an aperiodic task that generated packets following the Poisson distribution. Other experimental parameters are given in Table 2.

The compared methods are the greedy method, lazy method and optimal packet scheduling [8]. The greedy method focuses on the transmission throughput without considering residual energy. It tries to maximize the throughput as long as the battery power is available. On the contrary, the lazy method emphasizes the long lifetime of the system rather than the network throughput, delaying packet transmission as much as possible. Lastly, the optimal scheduling adjusts transmission periods so that the total transmission time of all packets may be minimized, based on accurate energy harvesting information. However, they did not propose any energy prediction method. In addition, they need to know entire packets that they will send in advance; thus, they cannot handle aperiodic packets generated irregularly.

Table 2. Simulation parameters.

Parameter	Value
Battery range (B^{min}, B^{max})	0, 10 MJ
The number of simulation	30
The duration of simulation	120 h
The rate of the periodic tasks (R1:R2:R3)	1:1.5:2
The energy per the periodic packet (E)	28 kJ
The deadline time of the aperiodic packet	5 min
T_s, T_f	19, 7 h

Figure 9 shows the variance of transmission numbers against the simulation time. At first, the greedy method sent more packets than the others, but the number of transmissions became stable during some intervals, e.g., 9 to 25 and 31 to 47 h from the starting time. The greedy method consumed all of the energy very quickly, and it could not send more packets until the battery was recharged. Solar energy cannot be harvested at night. Furthermore, the solar radiation in the early morning and in the late afternoon was not strong enough to harvest as much as the energy needed for packet transmission. On the other hand, the lazy method held off packet transmissions throughout the simulation so long as their deadlines had not passed. Conclusively, as time grew, the optimal and the proposed ACSE showed the best performance in terms of throughput. Note that, mentioned before, the optimal method needs several strict assumptions, e.g., entire packet generation information and accurate estimation of harvested energy.

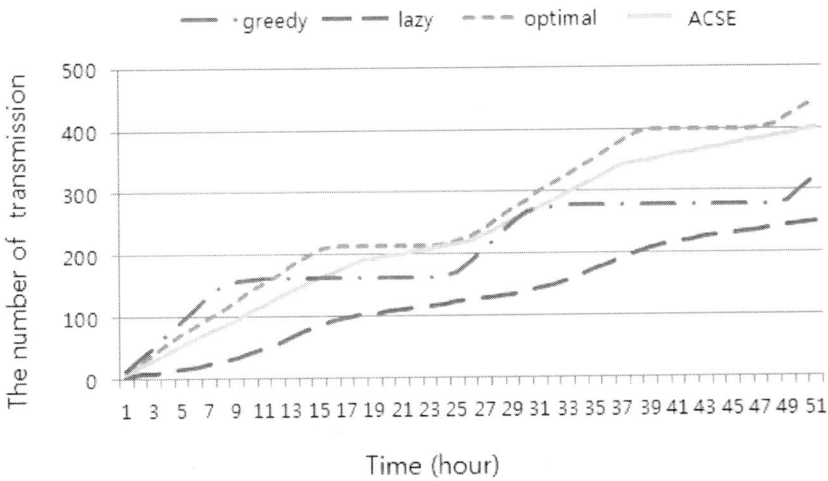

Figure 9. The number of packet transmissions.

Figure 10 illustrates the trace of the battery energy level change. The greedy method stayed at the complete discharge of the battery for a long time. The optimal method also often made the battery dissipate completely, since it is based on the assumption that the harvested energy can be accurately estimated. Our experiment showed an error of 6.78% compared with the actual harvested energy. On the other hand, the lazy method stayed at the full-charge of battery for a long time at the cost of low throughput. In this state, more solar energy is overflowed, which may have been stored and used by other methods. Unlike all other methods, ACSE always maintained the battery state between full charge and complete discharge. Thus, most aperiodic packets were successfully delivered even in the nighttime.

We observed the number of packets that missed the deadline due to the lack of energy. This experiment continued for five days and was repeated 30 times. As mentioned earlier, each sensor ran three periodic tasks and aperiodic tasks generated following the Poisson distribution. The arrival rates, λ, were set to 4, 8 and 12, and the results are given in Table 3. When $\lambda = 4$, ACSE missed the deadline of only three packets among a total of 491 packets, the success ratio being 99.4%. The greedy and optimal method could not satisfy the deadlines of packets frequently due to the battery dissipation, as shown in Figure 10. On the other hand, the lazy method satisfies the deadline best at the cost of throughput, since it pursues only the saving of energy for unexpected future use.

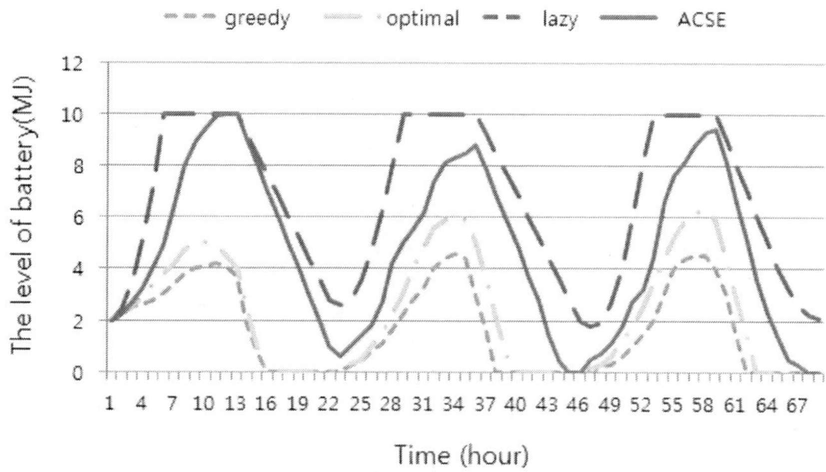

Figure 10. The change of the battery level.

Table 3. The number of packets missing the deadline. ACSE, adaptive control of packet transmission period with solar energy harvesting prediction.

λ	Greedy	Lazy	ACSE	Optimal
4	14	0	3	9
8	37	0	7	29
12	136	11	23	106

Besides the similar performance to the optimal scheduling in terms of throughput, ACSE also has a lower deadline miss ratio. These advantages are because ACSE dynamically controls transmission periods considering not only residual energy, but also the variance of harvested energy. Therefore, we argue that the proposed method ACSE can provide reasonable and realistic performance for energy harvesting wireless sensor network applications.

6. CONCLUSIONS

Our study suggests a novel estimation scheme for energy harvesting within three hours and ACSE for controlling packet transmissions according to the estimated energy. The estimation scheme utilizes weather factors, such as the past insolation and cloudiness. The scheme analyzes the correlation between these factors with consideration of the Earth's rotation. On the other hand, ACSE controls the packet transmission adaptively based on the estimated harvested energy. In our experiment, the integration of those schemes not only improved the prediction accuracy, but also enhanced the network throughput. Additionally, ACSE showed a lower deadline missing ratio and a higher energy utilization than the compared methods. Therefore, we can say that ACSE is an approach that can be used in a practical network. In the future, we have to handle two more issues: First, our prediction scheme for energy harvesting is relatively inaccurate for heavily cloudy days. Second, as mentioned earlier, the KMA data are measured at only several points in a city; thus, the past insolation and cloudiness might be a little different in the physical location of our sensor network. These errors can be corrected through data collection at the real position for a long time.

ACKNOWLEDGEMENTS

This research was supported by the Basic Science Research Program through the National Research Foundation of Korea (NRF) funded by the Ministry of Education (NRF-2013R1A1A2013120).

AUTHOR CONTRIBUTIONS

Kideok Kwon conceived the prediction method idea and implemented it. Jihoon Yang contributed to the basic concept of the system and evaluated our method. Younghwan Yoo gave some feedback as an advisor and discussed some problems about our research.

CONFLICTS OF INTEREST

The authors declare no conflict of interest.

REFERENCES

1. Sudevalayam, S.; Kulkarni, P. Energy Harvesting Sensor Nodes: Survey and Implications. *IEEE Commun. Surv. Tutor.***2011**, *3*, 443–461.

2. Gunduz, D.; Stamatiou, K.; Michelusi, N.; Zorzi, M. Designing Intelligent Energy Harvesting Communication Systems. *IEEE Commun. Mag.* **2014**, *1*, 210–216.

3. Sharma, N.; Gummeson, J.; Irwin, D.; Shenoy, P. Cloudy Computing: Leveraging Weather Forecasts in Energy Harvesting Sensor Systems. Proceedings of the Sensor Mesh and Ad Hoc Communication and Networks (SECON), Boston, MA, USA, 21–25 June 2010; pp. 1–9.

4. Edward, S.; Rogers, S. SEA-OR: Spectrum and Energy Aware Opportunistic Routing for Self-Powered Wireless Sensor Networks. Proceedings of the ACM Mobile Computing and Networking (MobiCom), Maui, HI, USA, 7–11 September 2014; pp. 429–432.

5. Hsu, J.; Zahedi, S.; Kansal, A.; Srivastava, M. Adaptive Duty Cylcing for Energy Harvesting Systems. Proceedings of the International Symposium on Low Power Electronics and Design, Tegernsee, Germany, 4–6 October 2006; pp. 180–185.

6. Moser, C.; Bruneeli, D.; Thiele, L.; Benini, L. Real-time Scheudling for Energy Harvesting Sensor Nodes. *Real-Time Syst.* **2007**, *3*, 233–260.

7. Manjeshwar, A.; Agrawal, D.P. APTEEN: A Hybrid Protocol for Efficient Routing and Comprehensive Information Retrieval in Wireless Sensor Networks. Proceedings of the International Paralles and Distributed Processing Symposium, Ft., Lauderdale, FL, USA, 15–19 April 2001.

8. Yang, J.; Ulukus, S. Optimal Packet Scheduling in an Energy Harvesting Communication System. *IEEE Trans. Commun.* **2012**, *1*, 220–230.

4

A SECURE AND ENERGY-BALANCED ROUTING SCHEME FOR MOBILE WIRELESS SENSOR NETWORK

Bei Liu, Yuanming Wu

School of Optoelectronic Information, University of Electronic Science and Technology, Chengdu, China

ABSTRACT

Mobile wireless sensor network (MWSN) has the features of self-organization, multiple-hop and limited energy resources. It is vulnerable to a wide set of security attacks, including those targeting the routing protocol functionality. In this paper, the existing security problems and solutions in MWSN are summarized, and then a trust management system based on neighbor monitoring is proposed. In the trust management system, the trust value is calculated by the neighbor monitoring mechanism, and the direct trust value and the indirect trust value are combined to establish the distributed trust model to detect the malicious nodes. The consistency check algorithm is capable of defending against the attacks on the trust model. In addition, because of the limited energy of the sensor nodes, the energy-balanced algorithm is introduced to prolong the lifespan of MWSN. The residual energy and energy density are considered in the routing decision. Finally, the simulation experiments show that the proposed algorithm can detect the malicious nodes effectively and achieve the energy-balanced goal to prolong the lifespan of MWSN.

KEYWORDS

Mobile Wireless Sensor Network, Network Security, Trust Model, Energy-Balanced

1. INTRODUCTION

Mobile wireless sensor network consists of a large number of micro sensor nodes deployed in the monitoring area, and forms a self-organized network system by the wireless multi-hop communication. Mobile wireless sensor network does not require a fixed infrastructure support, and is widely used in military reconnaissance, medical surveillance, environmental monitoring, agricultural breeding and other commercial areas [1] . Wireless sensor network has the following characteristics: 1) limited hardware resources; 2) limited power capacity; 3) no center; 4) self-organization; 5) multi-hop routing; 6) dynamic topology; 7) a large number of nodes and dense distribution. Based on the above characteristics, the design of mobile wireless sensor network which is a special network is challenging.

In data-centric MWSN, because of interference, collision and bad wireless channel, data in the transmission process may make mistakes or miss, resulting in being not correctly sent to the SINK node. The classic encryption and authentication mechanism can resist external attacks such as tapping data and breaking network communication, but can't solve the problem of internal attack. When the normal sensor node is captured and cracked by the enemy, it becomes a malicious node, attacks the network, and damages the normal function of the network. Table 1 lists common types of malicious node attacks.

Mobile wireless sensor network faces various threats. In [2] , a lightweight secure routing protocol is proposed, which can resist the known attacks that break the basic structure of the route. A protocol based on secure routing feedback is proposed [3] . Feedbacks from the neighbor nodes and base station provide the dynamic information of the current network. The secure multicast routing protocol is proposed in [4] . The trust model based on neighbor node monitoring is widely studied in recent years. A node evaluates its neighbor nodes and calculates the trust values as the decision parameters of the routing selection stage [5] . The method of obtaining trust information and calculating the trust value of each node is called trust model that can be classified by the different monitoring behaviors [6] . According to the distribution of trust decision-making mechanism, trust model can be divided into centralized [7] , hierarchical [8] or fully distributed [5] . There are a series of behaviors needing to be monitored. Each behavior corresponds to a trust metric. The trust metrics should be comprehensive, including data forwarding [5] , location verification [9] , data packet integrity and authenticity [8] , etc. The various communication behaviors need to be turned into trust values or varying levels of trust. Trust exchange mechanism is proposed [10] [11] to improve the reliability of trust information effectively. However, these mechanisms not only increase the resource consumption, but also make the trust model itself easily attacked. For example, a malicious node deceives the trust exchange mechanism through exchanging false trust information [12] . In the trust model based on direct observation [6] [7] [13] , the data that the node itself monitors are not entirely accurate. So the combination of the direct trust values and the indirect trust values obtained by the third party nodes is proposed [14] . This paper proposes a trust management approach from the human social relations to defend against malicious attacks. A node monitors the behaviors of its neighbor nodes to evaluate their trust values and then considers these trust values when the node selects the next hop forwarding node. The trust values can be directly obtained by the node, and can also be provided by other neighbor nodes.

In addition to the security threat caused by malicious node attacks, mobile wireless sensor network needs to consider the problem of limited node energy. The energy factor is considered in the routing algorithm to prolong the lifespan of the network. MDD (Multipath Directed Diffusion) routing algorithm expands the DD (Directed Diffusion) routing algorithm, and establishes multiple paths from the source node to the destination node to balance the energy consumption of each node [15] . The routing algorithm with multiple SINK nodes can reduce the paths of data transmission and save node energy [16] . In [17] , LEACH (Low Energy Adaptive Clustering Hierarchy) is proposed which adopts a hierarchical structure and selects cluster heads periodically. Since each node has the same opportunity to be chosen as the cluster head, the algorithm can balance the energy consumption of all nodes. In [18] , the EBRP (Energy-Balanced Routing Protocol) routing algorithm is proposed, which can achieve the energy-balanced effect based on the depth, residual energy and energy density. The routing algorithm in this paper considers the energy factor that includes the remaining energy and energy density to prolong the lifespan of the network.

Table 1. Malicious attacks.

Attack type	Attacker behavior
Selective forwarding, black-hole, grey-hole	A malicious node discards part or all packets
Sinkhole attack	A malicious node tries to attract traffic advertising and fake routing information, and then it does not forward it
Replay attack	A malicious node continues to send the same data repeatedly
Link spoofing attack	A malicious node can deceive the link layer verification mechanism, so that the sender of the packet thinks that the packet has been forwarded successfully
Modification attack	A malicious node modifies the data packets that it forwards
Sybil attack	An attacker presents multiple identities
Collusion attack	Many powerful attackers work in collusion to implement attacks
Flooding attack	A malicious node sends a large number of packets to overwhelm the normal nodes
Bad mouthing attack	A malicious node provides dishonest recommendations

In the rest of the paper, the related work is summarized in Section 2, while in Section 3 a secure and energy-balanced routing scheme is proposed in detail. In Section 4, the simulation experiments for the algorithm are carried out. Finally, conclusions are drawn in Section 5.

2. PROBLEM FORMULATION

2.1. Security Problem

In the mobile wireless sensor network, all nodes randomly move, the neighbor node set of each node is not fixed. A node monitors its neighbor nodes and evaluates their communication behaviors. When the node detects a malicious node from the neighbor nodes of the node, the node removes the malicious node from the neighbor nodes and no longer communicates with the malicious node. But the malicious node is able to move to a new location to continue to implement the attacks. So, after the node detects the malicious node, the node needs to notify the entire network of the ID of the malicious node. Considering a more complex situation, after the malicious node is detected, the malicious node changes its ID, and then moves to a new location to implement attacks. Because the ID of the malicious node is new, other nodes think that the malicious node is a normal node. As shown in Figure 1, this brings serious problems to the monitoring of malicious nodes. In this paper, the tracing technique through neighbor nodes is introduced to defend against this attack.

2.2. Limited Energy Problem

The weighted average of the depth, residual energy and energy density is introduced as the path decision criteria in EBRP. The algorithm can achieve energy-balanced effect in immobile wireless sensor network, but there are some problems in mobile wireless sensor network. In EBRP, the formula for calculating energy density is:

$$ED(i) = \sum_j \frac{E(j)}{S(i)} \tag{1}$$

The j means the neighbor node of the node i, E(j) is the residual energy of the node j, S(i) is the communication coverage area of the node i.

Due to node mobility, the neighbor nodes of the node i change constantly, if every time the neighbor nodes change, all values in Formula (1) are calculated again for a sum operation, this calculation is complex. Because every time the neighbor nodes change, the neighbor node set either adds a new node or reduces an old node. So the energy density only needs to add the value of the new node or subtract the value of the old node. The detailed formula is given in Chapter 3.4.

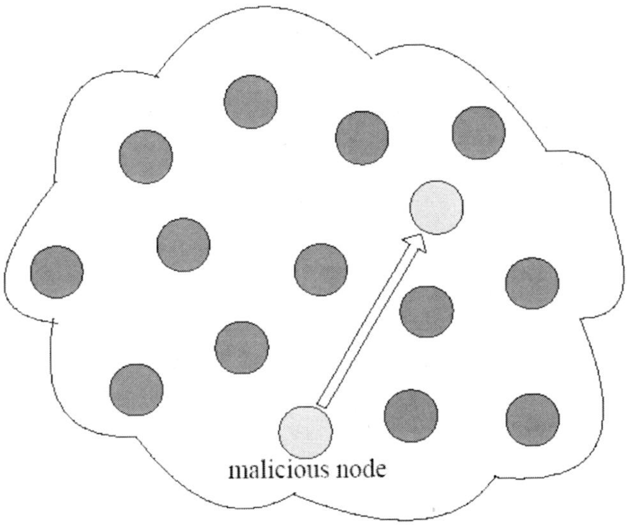

Figure 1. The malicious node moves and changes its ID.

In addition, Formula (1) for calculating energy density is problematic. When the node j selects the next hop node for forwarding, it selects the forwarding node from the nodes that their depths are less than or equal to the depth of the node j. These nodes are the parent nodes or sibling nodes of the node j. So the node j should not consider all the neighbor nodes in Formula (1), but only the parent nodes and sibling nodes.

3. SECURE AND ENERGY-BALANCED ROUTING ALGORITHM

In this chapter, combining the distributed trust model and the energy-balanced algorithm, a novel, easy to deploy, safe and effective mobile wireless sensor network routing algorithm is proposed. Based on the mechanism of neighbor monitoring and trust exchange, the trust model combines direct trust value with indirect trust value to detect malicious nodes. In addition, the residual energy and energy density of the nodes are considered in the routing decision to make the energy consumption balanced and prolong the lifespan of the network.

3.1. Network Topology

A large number of sensor nodes are distributed in a region randomly and uniformly to collect and forward data, which constitute a dynamic self-organizing network. The SINK node is located in the network to collect data and coordinate network.

3.2. Initialization and Update of the Depth

The SINK node broadcasts routing initialization message that includes message type and depth. The initial depth of each node is infinite. When a node receives the initialization message, if the depth of the message is greater than the depth of the node itself, the node updates the depth of the node itself and increases the depth of the message by 1, and then forwards the message. Eventually, each node has the correct depth, such as Figure 2.

Since the nodes are moving, the initial depths change in real time. There are 2 ways to update the depths. One is to update the depths by the SINK node in the global range like the process of route initialization. Because the depths of all nodes are updated, the energy consumption is great. The other one is that the nodes automatically update the depths. In the practical application, the moving distances of the nodes are limited in a short time, and the depths of most nodes are unchanged. When a node moves to a new location, the depth information can be obtained from neighbor nodes of the node. The node takes minimum depth from all depths of its neighbor nodes, and then increases the minimum depth by 1 as the depth of the node itself. As shown in Figure 2, the node 12 moves to a new place, and updates its depth.

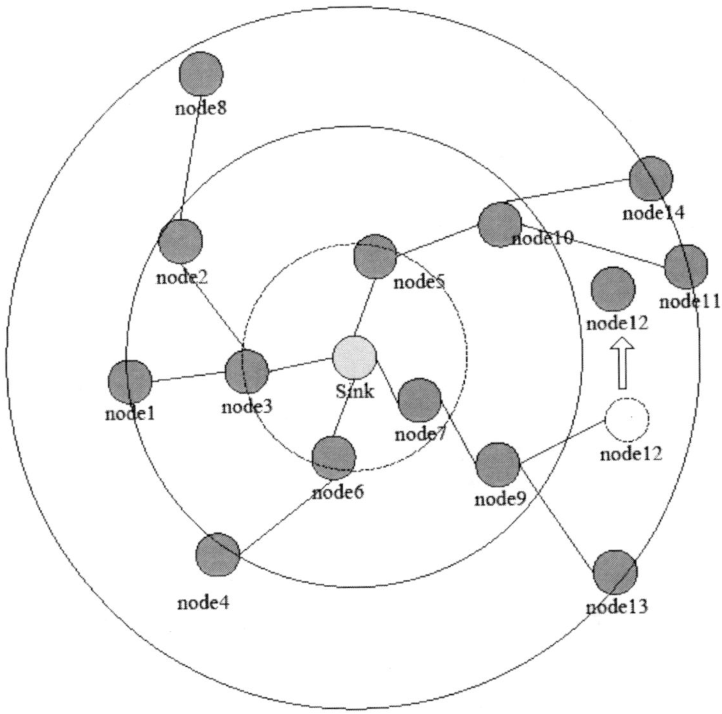

Figure 2. The depths of the nodes.

3.3. The Distributed Trust Model

The trust model consists of two modules that are the abnormal behavior detection module and the trust evaluation module. The abnormal behavior detection module establishes a series of abnormal behavior detection rules according to a variety of attack behavior features, and updates the statistics of normal and abnormal behaviors of the node according to the detection results. The trust evaluation module calculates the trust values according to the statistics, and judges whether the node is a malicious node or not. Table 2 is some symbol descriptions of this paper.

In the trust model, a set of behaviors for node detection are defined. Table 3 lists the trust metrics, each trust metric corresponding to one or a certain class of malicious attacks.

Specific detection methods are as follows:

1) Data forward: When the node i sends packets to the next hop node j, the node i is in promiscuous mode to overhear the wireless medium to ensure that whether the node j forwards the packets. In addition to the node i, the common neighbor nodes of the node i and the node j also monitor the node j.

2) ACK confirm: After the node i sends packets to the node j, the node i waits for the ACK from the node j. Considering the instability of the wireless communication channel, data retransmission mechanism is necessary, if the node i does not receive the ACK from the node j in a short time, the node i retransmits the packets. After trying the finite number of times, the node i gives up.

3) Data integrity: After the node i sends packets to the node j, the node i does not immediately delete the storage data of the packets, but monitors the node j and compares the forwarding packets of the node j with the storage packets of the node i.

4) Data flooding: If the number of packets received by the node i from the node j exceeds a certain threshold in a period of time, the behavior of the node j is considered abnormal.

5) Data duplication: If the number of the times that the node i receives same packet from the node j in a period of time exceeds a certain threshold, the behavior of the node j is considered abnormal.

6) Trust value response: The node i request trust information from the neighbor node j, if the neighbor node j responds the trust information to the node i, the number of successful times is increased by 1; otherwise the number of failed times is increased by 1.

7) Trust value verification: The trust values provided by the neighbor nodes need to be judged. This part will be given in the trust model.

The abnormal behavior detection module stores the detected results classified by the trust metrics. For a certain trust metric, if the behavior is normal, the α_{ij}^{Am} is increased by 1, otherwise the β_{ij}^{Am} is increased by 1.

Table 2. Symbol description.

Symbol	Description
i, j, k	Sensor node
A_m	Attack type, corresponding trust metric
R_{ij}	The distribution of the communication behavior of the node i about the node j, is a random variable
C_{ij}	The trust value of node i for node j, mathematical expectation for R_{ij}
R_{ij}^{Am}	For trust metric Am, the distribution of the communication behavior of the node i about the node j
C_{ij}^{Am}	For trust metric Am, the trust value of node i for node j, mathematical expectation for R_{ij}^{Am}
α_{ij}^{Am}	For trust metric Am, the number of nodes j normal communication behavior by node j monitor
β_{ij}^{Am}	For trust metric Am, the number of nodes j abnormal communication behavior by node j monitor

Table 3. Trust metrics for different malicious behaviors.

	Trust metric	Malicious behavior
1	Data forward	All types of packet loss attacks (black hole, gray hole, selective forwarding)
2	ACK confirm	The node does not confirm the ACK
3	Data integrity	Data tampering attack
4	Data flooding	Forward a large amount of data in a period of time
5	Data duplication	Delivery the same data repeatedly to the same node
6	Trust value response	In the trust exchange, the trust value is not returned to the requesting node
7	Trust value verification	"Badmouth" attack

There is a special case to consider: a malicious node performs well at the beginning to obtain the high trust value, and then implements attacks. The forgetting factor is introduced to detect the malicious node quickly. The trust value is calculated once every time period Δt, the formulas are as follows:

$$\alpha_{ij}^{Am} = \mu_1 \alpha_{ij}^{Am} + \alpha_{ij}^{Am}(\Delta t) \tag{2}$$

$$\beta_{ij}^{Am} = \mu_2 \beta_{ij}^{Am} + \beta_{ij}^{Am}(\Delta t) \tag{3}$$

$$\mu_1 < \mu_2 \tag{4}$$

μ_1 and μ_2 are the forgetting factor.

For the trust metric Am, the direct trust value is:

$$C_{ij}^{Am}\_direct = E\left(R_{ij}^{Am}\right) = E\left(Beta\left(\alpha_{ij}^{Am}+1, \beta_{ij}^{Am}+1\right)\right) = \frac{\alpha_{ij}^{Am}+1}{\alpha_{ij}^{Am}+\beta_{ij}^{Am}+1} \tag{5}$$

For all trust metric, the direct trust value is:

$$C_{ij}\_direct = \frac{1}{n}\left(C_{ij}^{A1} + \gamma C_{ij}^{A2} + \gamma^2 C_{ij}^{A3} + \cdots + \gamma^n C_{ij}^{An}\right) \tag{6}$$

The γ in the Formula (6) indicates the penalty coefficient for multiple attacks. Due to the penalty coefficient, when a malicious node launches a variety of attacks simultaneously, even though the n kinds of trust values are high, the comprehensive trust value is very low.

When $C_{ij}^{Am}\_total$ is below the trust threshold value θ^{Am} or C_{ij}_total is below the trust threshold value θ, the node i considers the node j as malicious node. The node i sends the information about the malicious node j to the SINK node and then the SINK node broadcasts the information to all nodes of the network.

After the malicious node is detected, the malicious node may change its ID, and then moves to a new location to implement attacks. The tracing technique through neighbor nodes is required to defend against this attack. In mobile wireless sensor network, the moving path of a node is continuous. A node will not suddenly appear in a new place. Based on the neighbor exchange mechanism, each node broadcasts the neighbor table to its neighbor nodes. In Figure 3, the node 3 moves to a new location, before and after the node 3 moves, the node 4 and 5 have always been the neighbors of the node 3. The node 4 and 5 are able to trace the mobile path of node 3 and then provide tracking information to the node 1 and 2. In this way, the node 3 is identified as a normal mobile node according to the neighbor node tracking information. If the node 3 is a malicious node, after it is detected, it changes the ID, and then moves to the new location. All neighbor nodes of the node 3 have no trace information about the node 3. So the neighbor nodes consider the node 3 as a malicious node.

3.4. Energy Factor

The node needs to consider the energy factor when making the routing decision. The energy factor includes the residual energy and energy density.

The detail energy model is shown in Figure 4.

Forwarding a k-bit message consumes energy as follows:

$$E_{Tx} = E_{Tx_{elec}}(k) + E_{Tx_{amp}}(k,d) = \begin{cases} kE_{elec} + k\varepsilon_{fs}d^2, & d < d_0 \\ kE_{elec} + k\varepsilon_{mp}d^4, & d \geq d_0 \end{cases} \tag{12}$$

To receive a k-bit message, the energy consumption is computed by:

$$E_{Rx} = E_{Rx_{elec}}(k) = kE_{elec} \tag{13}$$

In WSN, energy consumption is caused by communication and calculation, but communication energy consumption is the main. Therefore, Formula (12) and (13) will be used in simulation experiments to estimate the energy consumption of sensor nodes.

Formula (14) is to calculate the energy density of the node:

$$ED(i) = \sum\nolimits_j \frac{E(j)}{S(i)} \tag{14}$$

j means the parent nodes and sibling nodes that their depths are less than or equal to the depth of the node j, E(j) is the residual energy of the node j, S(i) is the communication coverage area of the node i.

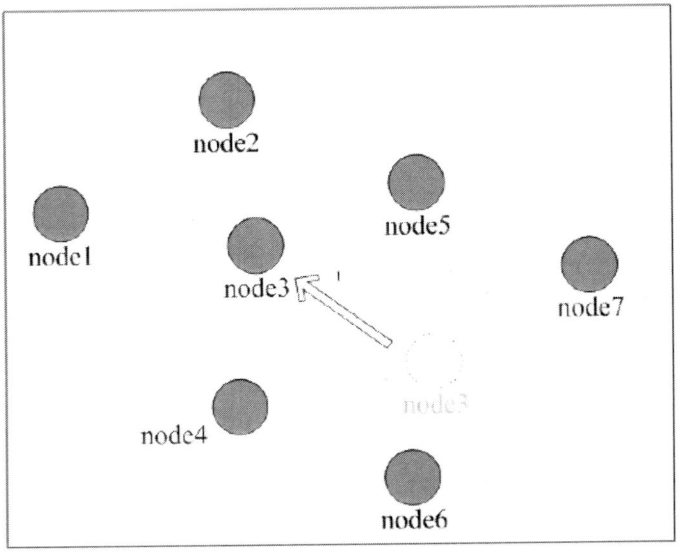

Figure 3. Tracing technique through neighbor nodes.

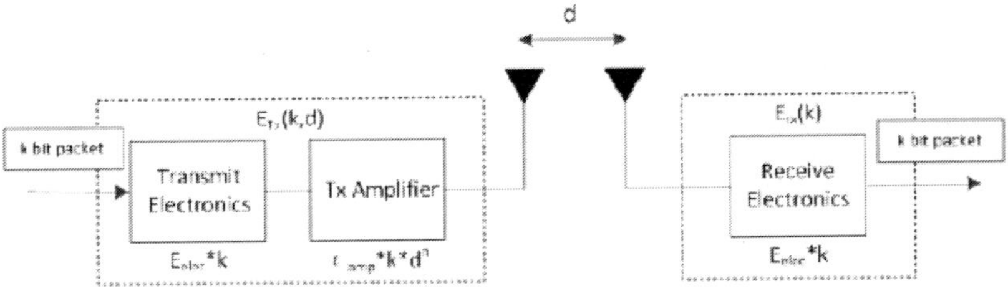

Figure 4. Energy model.

When the neighbor nodes of the node i change, the neighbor node set either adds a new node or reduces an old node. The energy density only needs to add the value of the new node k or subtract the value of the old node k in Formula (15).

$$ED(i)\_new = ED(i)\_old \pm \frac{E(k)}{S(i)} \tag{15}$$

The energy factor EE is calculated:

$$EE(j) = \alpha * E(j) + \beta * ED(j) \tag{16}$$

$EE(j)$ represents the energy factor of the node j. The coefficient α and β mean the proportion of the residual energy and energy density. The Chapter 3.5 will introduce the energy factor in detail in secure and energy-balanced routing algorithm.

3.5. Secure and Energy-Balanced Routing Algorithm

The proposed routing algorithm includes the depth update stage and the data transmission stage. The depth update stage is introduced in detailed in Chapter 3.2. In the data transmission stage the node makes the routing decision to choose the next hop node for forwarding data. When the node i forwards data, the routing decision is made according to the following steps:

Step 1: according to the trust model, the neighbor nodes of the node i are judged and the malicious nodes are removed.

Step 2: select the nodes that their depths are less than the depth of the node i.

Step 3: select the node that has the largest energy factor EE from the nodes that have been selected in the step 2. The selected node becomes the next hop forwarding node of the node i.

4. SIMULATION AND ANALYSIS

The experiments of this paper are carried on the simulation platform MATLAB. 200 nodes are randomly deployed in 400 m*400 m rectangular space, and the SINK node is in the center.

4.1. Initialization and Update of the Depth

The Figure 5 is the depth map after the initialization of the network, the circle represents the sensor node, and the square represents the SINK node. Some nodes are marked with a number that represents the depth of the node.

When a node moves randomly showed in Figure 6, the depth of the node is updated periodically showed in Figure 7. The depth of the node can be updated correctly and the change of the depth within a period is not more than 1.

4.2. The Distributed Trust Model

Combining direct trust value with indirect trust value, the trust model is simulated. 50 nodes are randomly selected from the network to send 6 packets to the SINK node per second. After 10 seconds 3 malicious nodes are randomly selected to implement selective forward attack. The packet loss rates respectively are 30%, 40% and 50%. After the node with 30% packet loss rate is detected, it changes the ID and moves to a new location to implement selective forward attack again. In Figure 8, the data arrival rate is lower than 100% from 11 seconds, which shows that there is packet loss attack in the network. Figure 9 shows the number of times of malicious node attack. Because from 11 seconds the malicious nodes start to implement selective forwarding attack, the number of times begins adding. After 31 seconds, the numbers of attacks do not increase, it shows that the malicious node changes the ID and moves to a new location, the malicious node is still identified as malicious node. Figure 10 shows the failure rate for malicious node detection. The first malicious node is detected in 17 seconds and second in 21 seconds and third in 31 seconds. When all malicious nodes are detected, the probability of miss is 0. The simulation results show that the trust model can detect the malicious nodes effectively.

4.3. Algorithm Simulation

Malicious nodes implement attacks like Chapter 4.2. The simulation compares the lifespan of 2 algorithms. The algorithm a does not consider the energy factor in the routing decision, only choosing the secure node with minimum depth. The algorithm b is the proposed algorithm in this paper. The result in Figure 11 shows that the algorithm b with the energy factor can significantly improve the lifespan of the network. Combined with Chapter 4.2, the proposed algorithm is secure and energy-balanced routing algorithm.

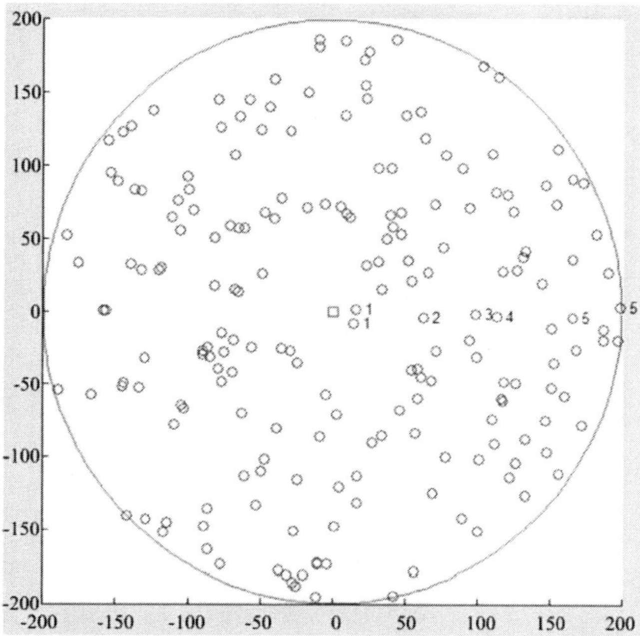

Figure 5. Initialization of the network.

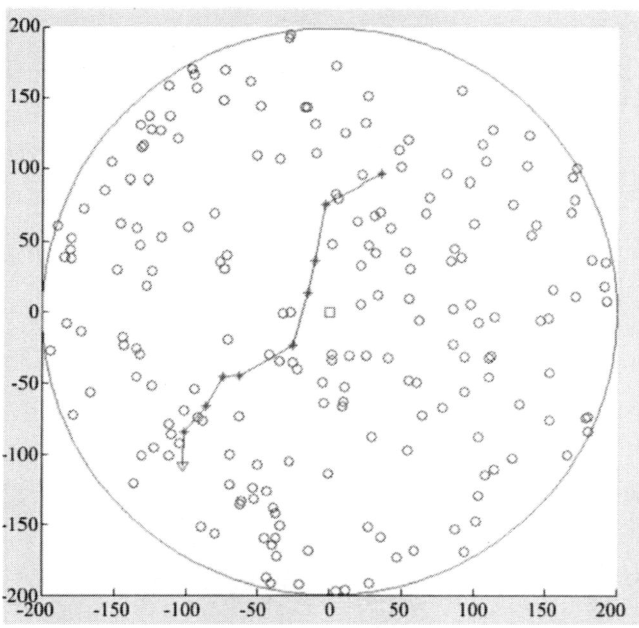

Figure 6. A node moves randomly.

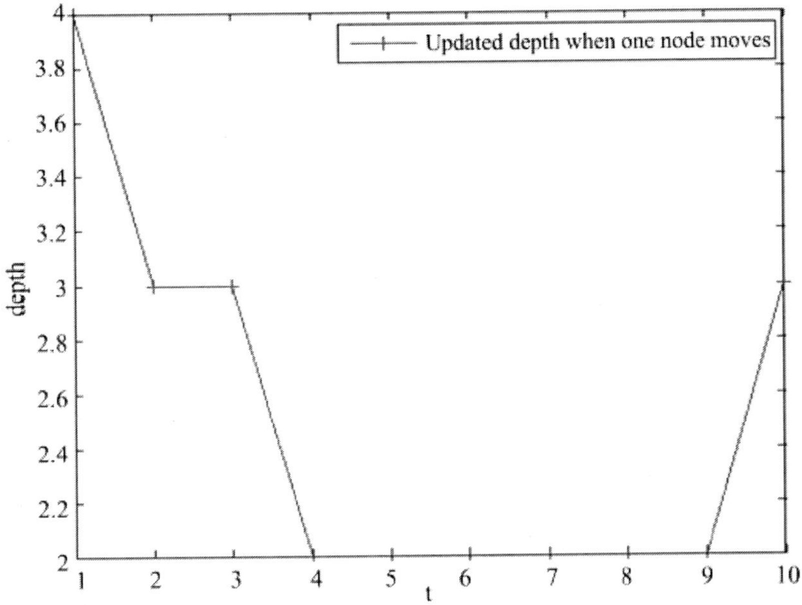

Figure 7. The depth of the node is updated.

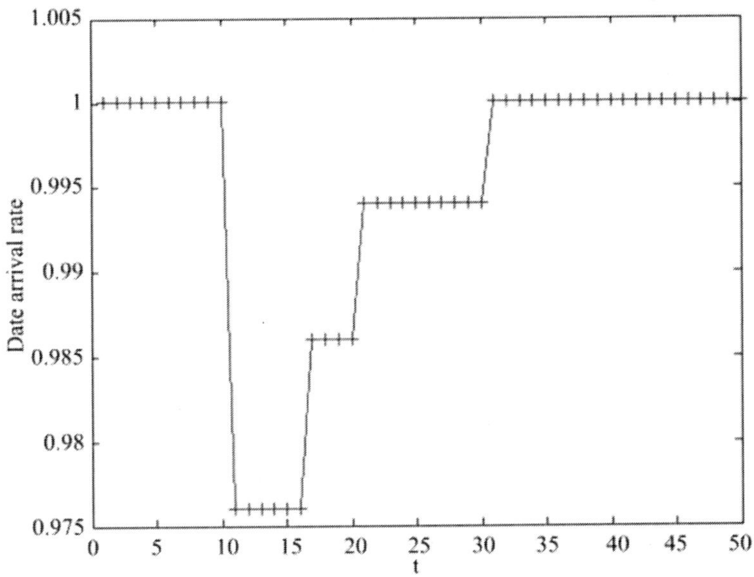

Figure 8. Packet arrival rate.

5. SECURE AND ENERGY-BALANCED ROUTING ALGORITHM

In this paper, a secure and energy-balanced routing algorithm which supports mobile wireless sensor network is proposed. The proposed algorithm adopts the trust model that realizes a distributed joint detection model combining the direct trust value with the indirect trust value. Meanwhile, the energy-balanced algorithm based on nodes' residual energy and energy density is introduced to prolong the lifespan of the network. Simulation results show that the routing algorithm is capable of detecting malicious nodes successfully. The consistency check algorithm of the indirect trust values helps the network to defend against the attacks on the trust model. The algorithm also prolongs the lifespan of the network.

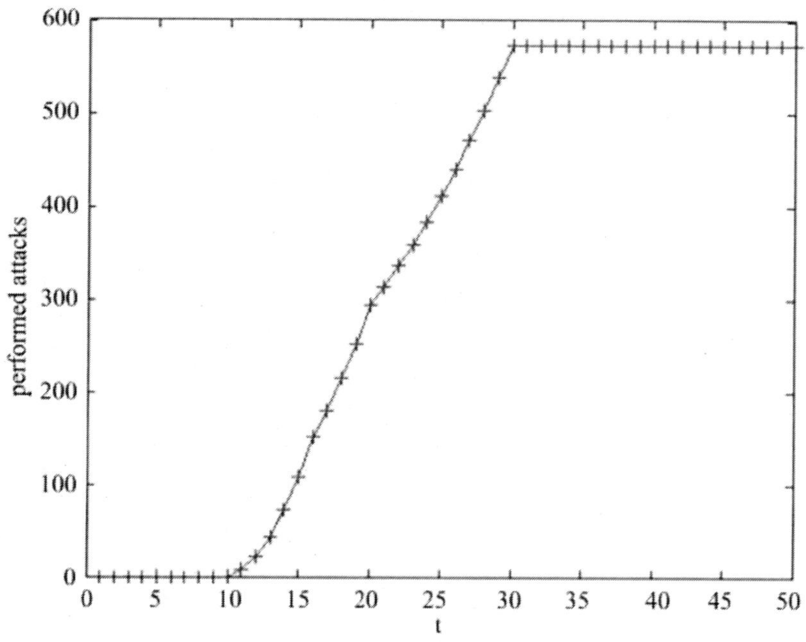

Figure 9. Total numbers of attacks.

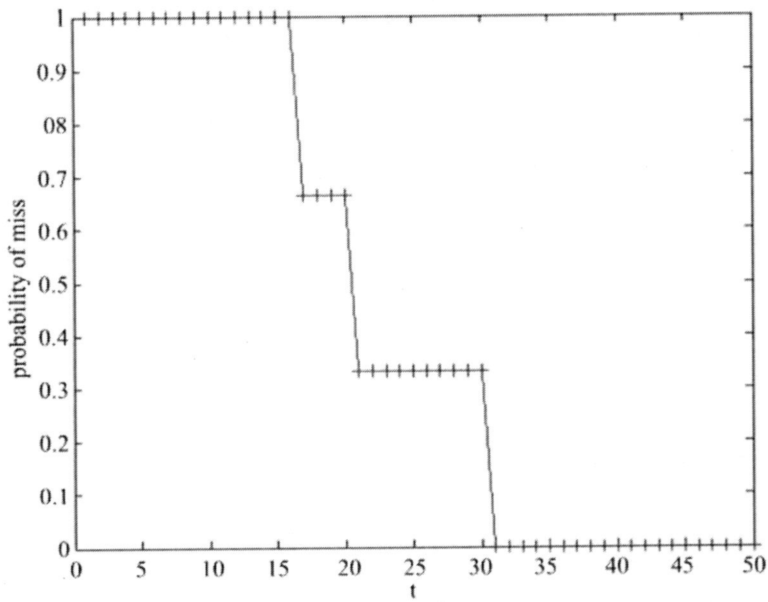

Figure 10. Probability of miss.

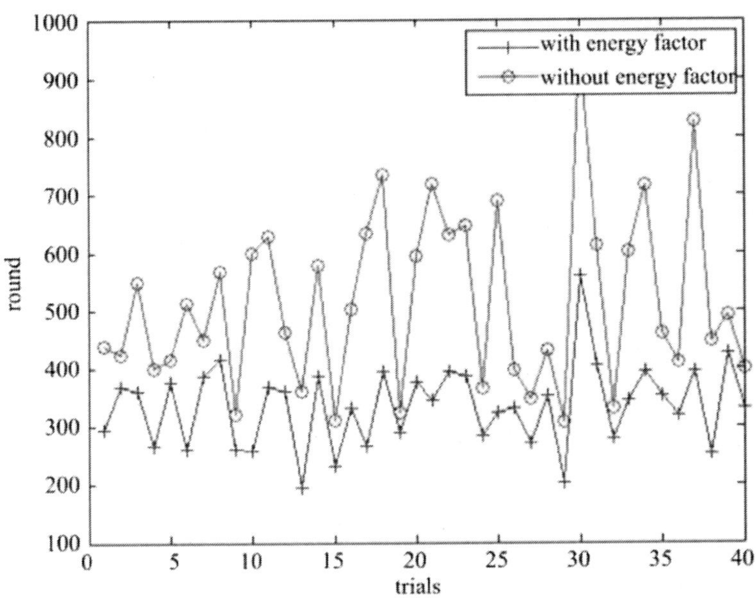

Figure 11. Lifespan of the network.

ACKNOWLEDGEMENTS

The authors wish to acknowledge the anonymous reviewers for their valuable comments and suggestions. We also thank MATLAB Technologies, Inc., for providing us with Wireless Modeler to validate our approaches.

REFERENCES

1. Akyildiz, I.F., Melodia, T. and Chowdury, K.R. (2007) Wireless Multimedia Sensor Networks: A Survey. IEEE Wireless Communications, 14, 32-39.

2. Deng, J., Han, R. and Mishra, S. (2006) INSENS: Intrusion-Tolerant Routing for Wireless Sensor Networks. Computer Communications, 29, 216-230.

3. Wood, A.D., Fang, L., Stankovic, J.A., et al. (2006) SIGF: A Family of Configurable, Secure Routing Protocols for Wireless Sensor Networks. Proceedings of the Fourth ACM Workshop on Security of Ad Hoc and Sensor Networks, Alexandria, 30 October 2006.

4. Dong, J., Curtmola, R. and Nita-Rotaru, C. (2011) Secure High-Throughput Multicast Routing in Wireless Mesh Networks. IEEE Transactions on Mobile Computing, 10, 653-668.

5. Hur, J., Lee, Y., Yoon, H., Choi, D. and Jin, S. (2005) Trust Evaluation Model for Wireless Sensor Networks. Proceedings of Advanced Communication Technology Conference, ICACT 2005, 1, 491-496.

6. Ganeriwal, S., Balzano, L.K. and Srivastava, M. (2008) Reputation-Based Framework for High Integrity Sensor Networks. ACM Transactions on Sensor Networks, 4, 3.

7. Crosby, G., Pissinou, N. and Makki, K. (2006) Location-Aware, Trust-Based Detection and Isolation of Compromised Nodes in Wireless Sensor Networks. International Journal of Network Security, 12, 107-117.

8. Theodorakopoulos, G. and Baras, J.S. (2006) On Trust Models and Trust Evaluation Metrics for Ad-Hoc Networks. IEEE Journal on Selected Areas in Communications (JSAC), 24, 318-328

9. Sun, Y., Yu, W., Han, Z. and Liu, K.J.R. (2006) Information Theoretic Framework of Trust Modeling and Evaluation for Ad Hoc Networks. IEEE JSAC Special Issue on Security in Wireless Ad Hoc Networks, 24, 305-317.

10. Li, H. and Singhal, M. (2006) A Secure Routing Protocol for Wireless Ad Hoc Networks. Proceedings of the 39th Hawaii International Conference on System Sciences.http://www.computer.org/csdl/proceedings/hicss/2006/2507/09/250790225a.pdf

11. Kang, K.D., Liu, K. and Abu-Ghazaleh, N. (2006) Securing Geographic Routing in Wireless Sensor Networks. 9th Annual NYS Cyber Security Conference: Symposium on Information Assurance, Albany, 14-15 June 2006, 38-49.http://www.albany.edu/iasymposium/proceedings/2006/kang.pdf

12. Akyildiz, I.F., et al. (2002) A Survey on Sensor Networks. IEEE Communications Magazine, 40, 102-114.

13. Lindsey, S. and Raghavendra, C.S. (2002) PEGASIS: Power-Efficient Gathering in Sensor Information Systems. IEEE Aerospace Conference Proceedings, 3, 1125-1130.

14. Hung, K.S., Lui, K.S. and Kwok, Y.K. (2007) A Trust-Based Geographical Routing Scheme in Sensor Networks. Proceedings of IEEE Wireless Communications and Networking Conference, Hongkong, 11-15 March 2007, 3123-3127.

15. Eghbali, A.N., Javan, N.T. and Dehghan, M. (2007) Load-Balancing Using Multi-Path Directed Diffusion in Wireless Sensor Networks. Lecture Note on Computer Science, 4864, 44-55

16. Cicirello, P., Mottola, L. and Picco, G.P. (2007) Efficient Routing from Multiple Sources to Multiple Sinks in Wireless Sensor Networks. Proceedings of the 4th European Conference on Wireless Sensor Networks (EWSN 2007), Delft, 29-31 January 2007, 34-50.

17. Heinzelman, W., Chandrakasan, A. and Balakrishnan, H. (2000) Energy-Efficient Communication Protocol for Wireless Microsensor Networks. Proceedings of the 33rd Hawaii International Conference on System Sciences (HICSS'00), 4-7 January 2000.

18. Ren, F.Y., et al. (2011) EBRP: Energy-Balanced Routing Protocol for Data Gathering in Wireless Sensor Networks. IEEE Transactions on Parallel and Distributed Systems, 22, 2108-2125.

5

CONTEXT AWARE WIRELESS SENSOR NETWORK SUITABLE FOR PRECISION AGRICULTURE

Nour Brinis, Leila Azouz Saidane

National School of Computer Science, University of Manouba, Manouba, Tunisia

ABSTRACT

Using the Wireless Sensor Networks WSNs in a wide variety of applications is currently considered one of the most challenging solutions. For instance, this technology has evolved the agriculture field, with the precision agriculture challenge. In fact, the cost of sensors and communication infrastructure continuously trend down as long as the technological advances. So, more growers dare to implement WSN for their crops. This technology has drawn substantial interests by improving agriculture productivity. The idea consists of deploying a number of sensors in a given agricultural parcel in order to monitor the land and crop conditions. These readings help the farmer to make the right inputs at the right moment. In this paper, we propose a complete solution for gathering different type of data from variable fields of a large agricultural parcel. In fact, with the in-field variability, adopting a unique data gathering solution for all kinds of fields reveals an inconvenient approach. Besides, as a fault-tolerant application, precision agriculture does not require a high precision value of sensed data. So, our approach deals with a context aware data gathering strategy. In other words, depending on a defined context for the monitored field, the data collector will decide the data gathering strategy to follow. We prove that this approach improves considerably the lifetime of the application.

KEYWORDS

Wireless Sensor Network, Precision Agriculture, Data Collector, Context Aware

1. INTRODUCTION

Water consumption in agriculture is the most part of water resources consumption in domestic, industrial and agricultural purposes. So, valuation of agricultural water productivity is important for improving the agricultural water resources utilization, especially in the arid areas [1] . For this reason, the science of precision agriculture has been involved, having the purpose to enhance efficiency in crop farming and water resources. This science consists of precise understanding, estimating and evaluating crops condition with the aim of different parameter readings. WSN (Wireless Sensor Networks) is one of the promising technologies for monitoring crops and doing the suitable readings. A number of studies have been published in the last few years on this theme [2] -[5] . We distinguish between two kinds of applications: reactive WSN applications and data gathering applications. Using a reactive WSN, even the person in the field is replaced by an automatic actuator to do some critical tasks like irrigation and fertilization when sensed parameters achieve a given value; this is called the Wireless Sensor Actuator Network WSAN. In the other case, the actuation task is not considered. So, the WSN has the limited purpose of monitoring the field. In case of reactive WSN, the precision farming system has the following parts:

1) Identification of the sensing agricultural parameters.

2) Identification of the sensing location and deployment of the sensor network.

3) Transferring data from crop field to control station for decision making.

4) Actuation and Control decision based on sensed data.

In the second kind of applications, the actuation part is not considered. These parts are detailed in the following paragraphs.

2. A WSN-BASED PRECISION FARMING SYSTEM

2.1. Identification of the Sensing Agricultural Parameters

This task is performed by the agricultural engineer. It consists of choosing the agricultural parameters to sense in order to achieve the target. Some of the common and most critical measured factors in agricultural WSN are air temperature (T) and relative humidity (RH), soil moisture, salinity, etc. [6] [7] . Of course the decision of what factors to monitor depends on the crop type. In addition to these, there are other factors to measure such as leaf thickness and atmospheric pressure, which depends on the crop being monitored. The option of the sensor types such as Sensirion SHT75 or SHT71 for RH and T depends on factors related to sensor accuracy, resolution, range, power consumption, precision, cost, etc. Figure 1 present different sensors used for agriculture field. We notice that some platforms like Mica2 support many kind of sensor boards in order to do different readings at the same time [8] .

2.2. Sensor Network Deployment

Every deployment has its own needs imposed by the type of the monitored crop or plant or by other special application. The coverage of the sensor nodes in agricultural WSN must be dense, i.e. one sensor node every one square meter (1 m^2) [9] [10] . Dense deployment serves to capture all the necessary measurements in order to have complete and reliable knowledge of the monitored area. For instance, crops have always different climate, known as microclimate. The micrometeorological parameters outside the field are monitored with the installation of weather stations. Besides the aforementioned issues, precision agriculture has maintenance-safety issues. In fact, maintenance of the WSN system must be considered, because of the long-term nature of the deployment. The sensor nodes must be put into protective cases preventing them from moisture, mud, etc. These cases have ratings in form of IP00, which means no protection. The first digit means protection against solid objects, while the second one means protection against liquids and every level has its definition. For example, in IP67, the 6 digit means total protection against dust and the 7 digit means protection against the effects of temporal immersion till 1 m underwater [11] .

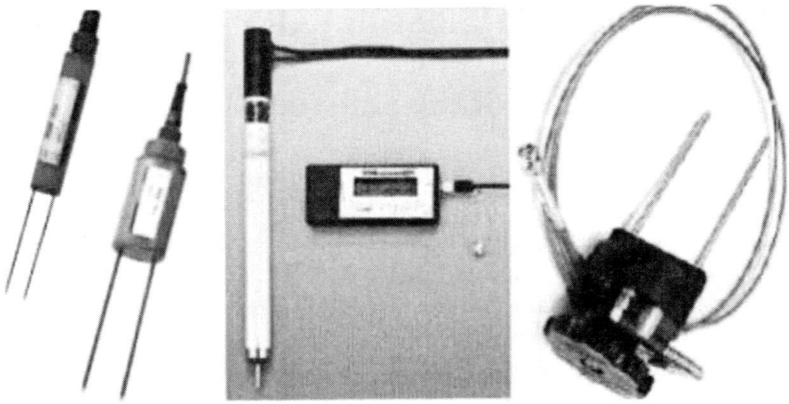

Figure 1. Agricultural sensors.

2.3. Transferring Data from Crop Field to Control Station

Many agricultural deployment strategies assume the sensor nodes sensing and sending data to be time-driven, such as in [12] -[14] , in order to acquire a complete picture of the crop circumstances and act accordingly. There are other sensing strategies like event-driven, which are mostly used when monitoring phenomena such as volcanoes, earthquakes, etc. In these cases the monitored subject is the event itself. Finally, the sensing task can be done on-demand based on the user/application requirements. The sensing intervals vary from one minute to one hour.

2.4. Actuation and Control Decision

Actuators like sprinklers, foggers, and valve controlled irrigation system are available and are controlled by the base station.

As a delay tolerant application, precision agriculture may rely on a data gathering strategy better than a tree- routing technique. In particular, we are interested in designing data gathering protocols in WSNs to increase the lifetime of the system, making it more energy efficient. Many data gathering schemes have been proposed for wireless sensor networks. Our data gathering strategy inspires from the strategies presented in the following paragraphs.

3. DATA GATHERING SCHEMES

3.1. EDGM: A Single Hop Gathering Strategy Using Data Mules

The goal of this strategy [14] is to determine the optimal number of data mules required to optimize the covered network lifetime while meeting a given data Delivery Delay Deadline DDD. The goal of each sensor node is restricted to capture data and then transmit them to a one-hop node: directly to the sink, if this node is located in its transmission range, or to a Data Mule DM, when it is one hop away. In other words, sensor nodes do not relay the traffic generated by other nodes. At the proximity of each static node, a DM stops at a Break Point, denoted BPoint, for a defined Break Period, denoted BPeriod, and sends a HELLO message. When it arrives at the proximity of the sink and the energy terminal, a DM sends the collected data to the sink. Each sensor node SN is associated to a unique BPoint. The break point positions are defined as the tops of equilateral triangles of edge length $\sqrt{3}rc$ with rc is the communication range of sensor nodes [15] . Each SN remains awake only during its break period to send its captured data. Then, it turns off its communication unit. During the break period, since several sensor nodes are authorized to send data to the DM, collisions may occur. A deterministic medium access scheduling is set by the DM, in order to avoid the collision problem. More precisely, each node starts sending its sensed data when it receives an invitation delivered by a DM through an INVIT message which contains the medium access scheduling. During the first tour, each SN receives the information needed to synchronize itself with the DMs present in the considered area. Once it is synchronized, it keeps awake only during a DM break period in its associated BPoint. During this time, it is authorized to send its collected data through an INVIT message. Figure 2 presents the data collector trajectory and break points.

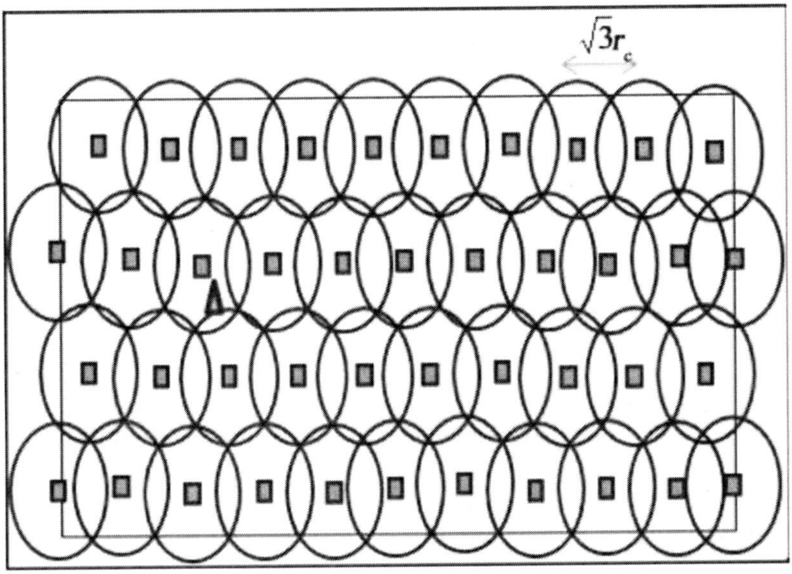

Figure 2. EDGM strategy.

3.2. ADA-DG: Adaptative Delay Aware Data Gathering Strategy

The key idea of ADA-DG [16] consists of finding the suitable trajectory of the Data Collector DC for which we guarantee sending data to the sink before a required data delivery delay deadline. For this purpose, each static sensor node sends their collected data to the DC with incorporating the tolerated data delivery delay deadline DDD. Knowing the DDD, the DC selects the suitable trajectory that allows carrying data to the sink at the appropriate time while minimizing as much as possible the static nodes energy consumption. The suitable trajectory is selected according to a suitable selection of a relay hops bound, denoted Hmax. Therefore, a subset of the whole break points is selected and it is colored Red, denoted RBPs. Only the RBPs must be visited by the DC. The remaining nodes attached to the remaining break points, colored green, will be affiliated to one of the RBPs. So, ADA-DG strategy aims at coloring the sensor network area into red-green disks. This problem is resolved by two steps. The _first step consists of determining the red breakpoints, at which the DC stops to gather data. This sub-problem is converted into a logical satisfiability problem. The propositional variable are the set of breakpoints {BP1, BP2, ..., BPn} where Bpi is set true means that the corresponding breakpoint is red. Knowing, the network graph, we set n logical clauses corresponding to the possibilities of each breakpoint to be colored red. For instance, let Ci be the clause corresponding to the break point BPi:

Ci: BPj or (BPk and BPl)

this constraint means that:

"BPi is Green"→ "BPj is red" or "BPk and BPl are red".

Our goal consists of satisfying all the clauses Ci. By resolving this problem, the different truth statement defines the different possibilities to construct red-green network. The goal of ADA-DG strategy is to find the trajectory visiting the minimum number of red break points; this means that the required solutions is that having the minimum number of truth values in their truth statement. We may have many solutions. So, the second criterion for selecting the RBP consists of minimizing the cost of the selected trajectory, in terms of number of relay hops. Therefore, [16] computes this cost for each truth solution. Then, the word offering the minimal cost is chosen to select the RBPs.

At the second step, the DC computes the shortest possible route that visits each RBP exactly once and turns back to the RBP containing the sink. This route is deducted using an heuristic for the Traveling Salesman Problem TSP.

Figure 3 presents an example of sensor network coloring using the ADA-DG strategy.

Our contribution in this paper consists of including context awareness in data gathering schemes for considering the in-field variability agricultural parcel. In fact, context awareness would incorporate a degree of intelligence in the sensor nodes so that they can decide the action to do, depending on a local context. Since we are dealing with agriculture, two types of contexts may be considered: the first one consists of the sensed data value such as temperature or humidity (current context) and the second one is an implicit input defined for different land conditions such as crop types, location or time of the day.

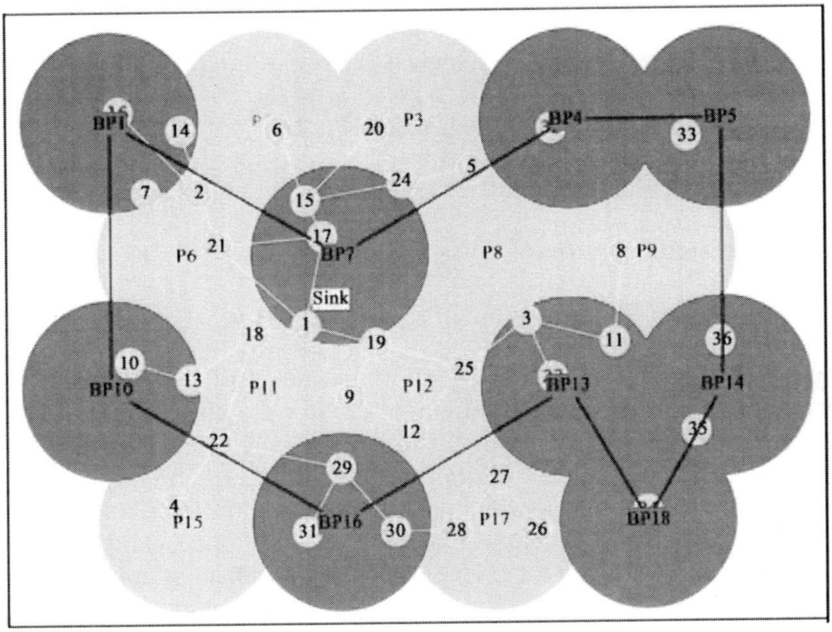

Figure 3. ADA-DG strategy.

4. CONTEXT-AWARE DATA GATHERING STRATEGY

In this paper, we deal about the topology illustrated in Figure 4. An agricultural parcel is divided into a number of fields, each of them are planted with a type of crop. We suppose that the data collector makes a smart decision about its trajectory depending on each field context. This idea tends to get over supporting a fixed kind of sensor topology for variable fields. In the two following section, we propose two data gathering approaches. Each of them uses a context definition in order to decide the data collector strategy.

4.1. Data Gathering Scheme: Based on ADA-DG Strategy

We now explain how we incorporate context awareness in choosing the data gathering strategy. Our approach is based on the ADA-DG strategy. Besides, we suppose that a Field Master FM is selected in each field. His role consists of saving the last context of the field and sending it to the Parcel Master PM using WIFI technology. The field master saves the worst value of captured data, adds the date of capture and sends them to the PM. We also note that the data gathering may be done by a data collector robot or any other instrument that visits the field to make insecticide, to harvest crops or any other reason. The decision of the strategy is made only by the PM after collecting the n last contexts. The PM will visit k fields ($k \leq n$) depending on the time of the last data gathering of each field, the last context and the required delay. Concerning the context model, we inspire from the pyramid model proposed in [16] . The agriculture information space is represented as a concept pyramid with the vertices representing the various contexts of information retrieval (temperature, humidity, land type, crop type). In our methodology, we consider a space level query for the whole agricultural parcel (crop, land type, temperature) and an edge level query for each field (land type, temperature). The FM saves the worst context which is defined as the worst values of sensed data (temperature). We denote DDDj the required Data Delivery Delay of field j from the k fields to visit. The fields are organized in croissant order of the DDD. These fields will be visited by the DC under control of the PM. The control is done by the PM depending on the last received context of each field and their DDD. We apply the ADA-DG strategy in each field. In other words, the DC goal consists of applying the pre-explained ADA-DG strategy for each part of the field. The DDD of each part is decided by the PM based on the last context of each part. Algorithm 1 resumes the behavior of the DC when receiving the decision from the base station. Figure 5 illustrates an example of this data gathering scheme application.

4.2. Data Gathering Scheme: Based on Islands Creation

In the first approach, we suppose that all the sensors deployed in a field send their data separately to the data collector. However, as a fault-tolerant application, precision agriculture does not require a high precision value of sensed data. So, in order to improve the lifetime of the application, we propose to group the sensor nodes into islands with near sensed values. For instance, in a given land, we may have some permanent variabilities, for example, a hill or plain space, a shadow or light area for a given time-of the day... So, having the same permanent condition of a given part of the field may imply finding very near values of sensed data. Then, sensing data from only one node of a part of the field, having near conditions, may satisfy the precision agriculture application. For this purpose, we propose in this section to group the nodes into islands having near sensed values. In the following paragraph, we define the islands definition. Once the islands are defined, alternatively each node is visited in the same time-of the day. The other nodes sleep in order to preserve their energy. We notice that a trigger of new island creation is defined. In our case, a high difference between the actual and the last sensed data from the same island triggers new islands creation. Other triggers may be defined depending on the type of crops. In the following paragraphs, we explain the island creation algorithm and the corresponding data gathering approach.

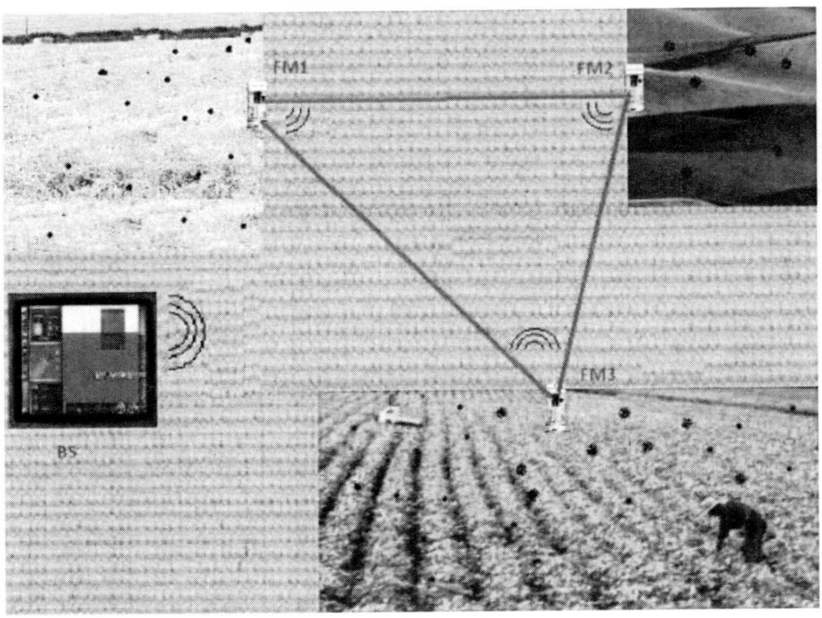

Figure 4. Parcel topology.

4.2.1. Island Creation Algorithm

The goal of this algorithm consists of grouping the sensor nodes having usually near values of sensed data into the same island. For this purpose, deployed sensors exchange the sensed data and compare them with the received values of their neighbors. Neighbor nodes having sensed data, that values are in a given precision interval [Vmin, Vmax], are considered in the same island. Otherwise, a new island is defined when a neighbor node senses data that values are out of the precision interval. We suppose that network is connected, which means that each node can be reached. The neighboring nodes exchange the following messages:

- CMsg (Check Message): containing the dynamic sensed data (such as temperature T), the neighbor nodes check the similarity of T with their sensed data (belonging to a given [Vmin, Vmax] interval).

- NIMsg (New Island Message): Announce a new island definition.

- EMsg (End Message): allow the Island Master IM to know the node's identifiers of its island.

Algorithm 1: DC moving path
INPUT: the network graph
OUTPUT: the DC moving path

1: j: the counter of fields
2. When arriving to a *FM*, update *D*.
    D is the time elapsed from the *PM* decision make until reaching the concerning *FM*
3. $DDD_j = DDD_j - D$
4. ADA-DG(DDD_j)
5. k is the counter of RBP
6. for each BP_k in RBP do
7.   collect data from the associated nodes
8. end for
9.   move to the next *FM*

Algorithm 1. DC moving path.

We denote:

- Di: the data sensed by node i.

- Ti: the instant when node i receives a CMsg.

- LTi: the last time when node i received a CMsg.

- Delay: the maximum delay to construct an island.

- Dbt: the minimum delay between two triggers to reconstruct islands.
- i is an end node if it has not descendant neighbors.
- CMsg (n, E, D): a CMsg sent by node n, passing by nodes in the set E. E is initially an empty set. D is the data captured by node n.

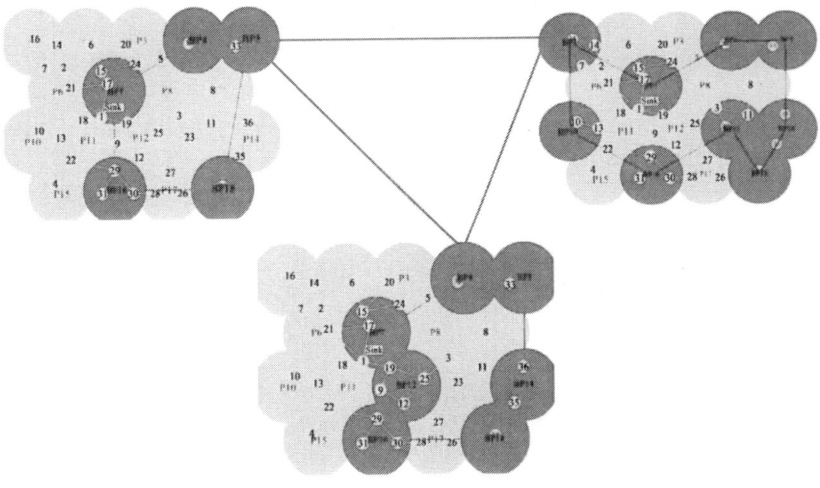

Figure 5. Example: data gathering based on AD-DG strategy.

The flowchart of the proposed algorithm is shown in Figure 6. The first node triggered to launch the algorithm, is declared IM. It sends a CMsg (IM, TIM) containing its ID and the sensed data to its neighbors. Each node j, receiving a CMsg, compares the received data with that sensed. In case of belonging to the precision interval, then if node j is neither an end node nor a node in a loop situation, than it sends the data to its neighbors after adding its ID. Otherwise, it sends an EMsg to the parent node (source node). In the other case, a NIMsg is sent to parent node.

An alternative circumstance could occur when more than one IM is declared in the process. This situation is marked by receiving more than one CMsg from different source nodes within a Dbt delay. In that case, the node waits until receiving all CMsgs, and then it send an EMsg to the IM, having the smallest ID, after including all nodes known from other received CMsgs.

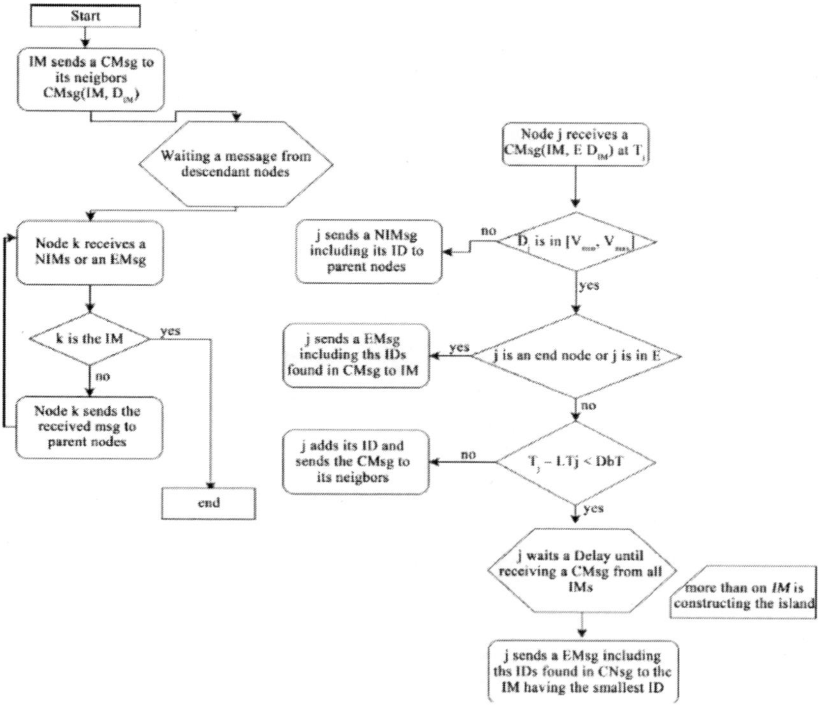

Figure 6. Flowchart of the island creation algorithm.

We notice that a node i concludes that it has received a CMsg (E1, T) in a loop situation when the last received CMsg (E2, T) is such that $E1 \subset E2$. After performing the algorithm, each IM maintains the IDs of nodes in its island and the IMs of neighbor islands.

4.2.2. Island Creation Example

For the sake of clarity, an example of islands creation with 3 distinct temperature intervals: (27) (29) (30) (32) and (33) (35) is illustrated in Figure 7.

Initially, the node 1 sends a CMsg including its identifier and the sensed temperature (30) to its neighbors. Each node receiving CMsg checks if its sensed data is in the corresponding temperature interval. In that case, it adds its identifier and forwards the CMsg to its neighbors. Otherwise, it sends a NIMsg to its parent nodes. Nodes receiving a

NIMsg forward it to the IM. In this case, node 9 senses a data that value is out of the interval (30) (33). So, it is declared IM and sends a NIMsg to the source node (node 3). This one forwards the message until the IM (node 1). Node 9 does not have descendant node. So, it forms an island containing only itself (node 9). Node 6 receives a CMsg in a loop situation, and then it will send an EMsg to IM (node 4).

4.2.3. Data Gathering Scheme

In this paper, we are interested to precision agriculture applications. In that case, network is not necessarily connected. However, knowing that the DC maintains the positions of all deployed nodes, we apply the island creation algorithm to each connected part of the network. Then, the DC will visit the different nodes that are declared IM. The first tour of the DC is a discovery tour. In the issue, the DC will maintain the real IM's list in an ascendant order of their identifiers and the identifiers of nodes composing each island.

The DC will visit only one sensor of each defined island IM. In order to balance the energy consumption between all nodes, IM is alternatively selected from island nodes and with the condition that at the same time of the day different IM are selected. Knowing the IM in each tour, a node can select its tour to be an IM. During the other tours, it remains in sleeping mode in order to conserve its energy. The reconstruction of islands is triggered by the DC.

5. SIMULATIONS AND RESULTS

In this section, we present simulation results to evaluate the efficiency of the proposed data gathering strategies. We carried out several simulations using NS2 simulator.

5.1. Simulation Parameters

Table 1 summarizes the simulation networking parameters.

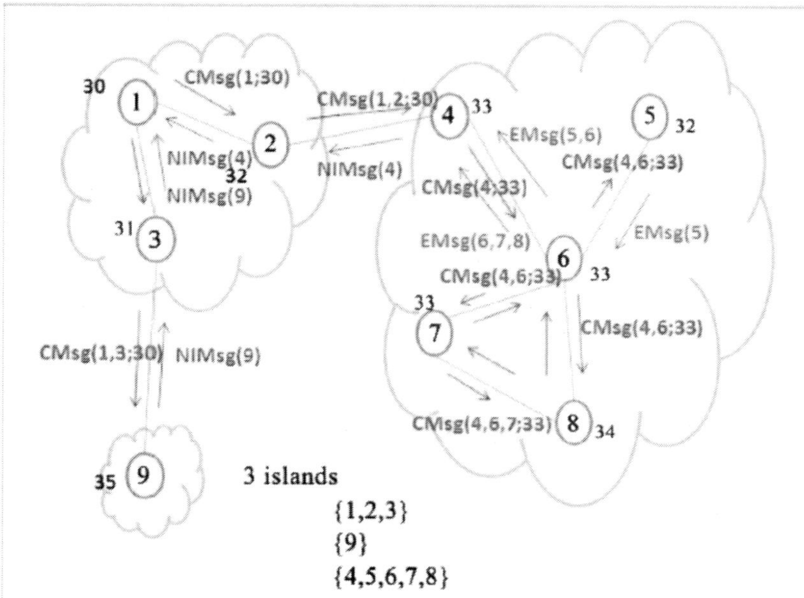

Figure 7. Example of island creation.

5.2. Simulation Results

We analyze the strategies efficiency in the two following case:

- case 1: using the first approach for which we suppose that readings are done every 1 hour when the temperature goes on or above the threshold level, here it is 30 degree Celsius. Otherwise, readings are done every 2 hours.

- case 2: using the first approach, readings are done every 1 hour whatever the previous sensed temperature.

In Figure 8, we plot the average residual energy. We observe a minimum gain of 25% of energy in the first case. So, when introducing context awareness based on the dynamic parameter temperature, we can gain in terms of network lifetime. In the first case, the DC can do more than 10 tours before sensors failure; however in the second case only 6 tours are done.

Table 1. Simulation parameters.

Network configuration	Number of fields	3
	Fields area	(200 m × 200 m), (300, 300), (400, 400)
	Rc	50 m
	Bandwidth	2 Mbps
	Mac protocol	802.11
	DC speed	1 m/s
Energy model	Type	Battery
	Initial energy	100 Joules
	Transmit power	0.36 watt
	Receive power	0.24 watt
	Sensing power	0.015 watt

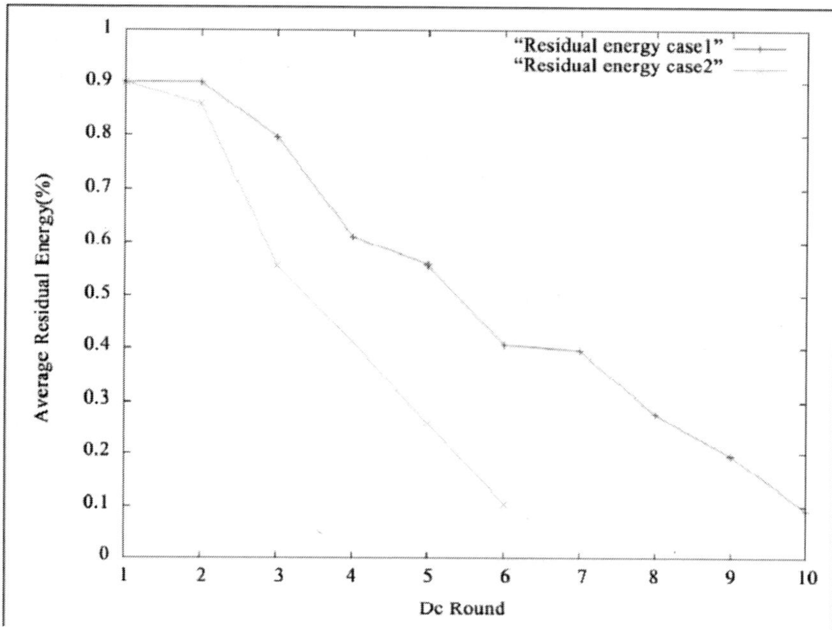

Figure 8. Average residual energy—first approach.

Similarly, in Figure 9, we compare the first approach case 2 and the second approach using readings every 1 hour. We observe that the second approach is more relevant with making sleep some nodes of a given field. In fact, only one sensor of the island, IM, must sense and send the sensed data to the DC, the other island nodes sleep.

In this section, we evaluate the effect of the node islands creation on the efficiency of the solution.Figure 10 presents the energy consumed rate after 4, 8 and 12 tours. Besides, we present the energy consumed for island creation. Obviously, we found that the energy increase with the number of tours. Moreover, when the number of groups increases, the corresponding number of island masters increase. So, we observe that the energy consumed will increase with the number of active nodes representing the island master nodes.

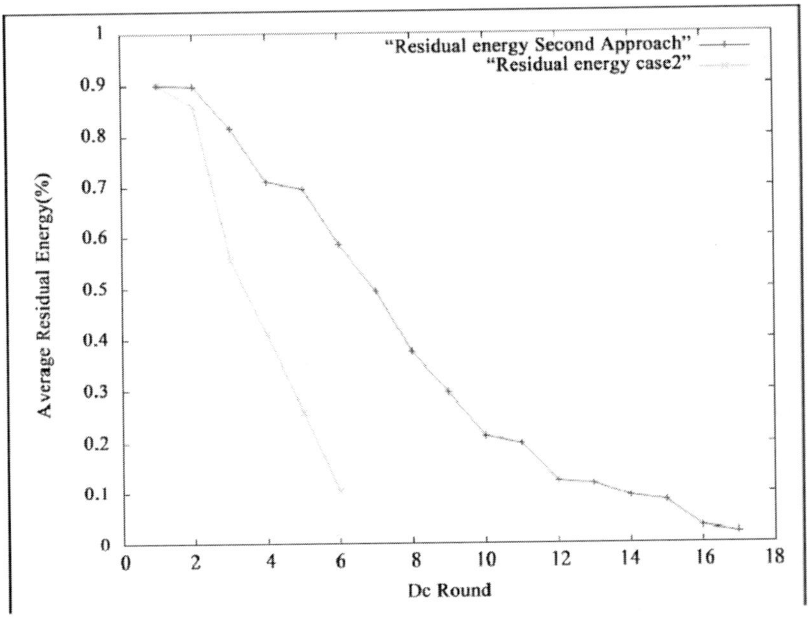

Figure 9. Average residual energy—first and second approaches.

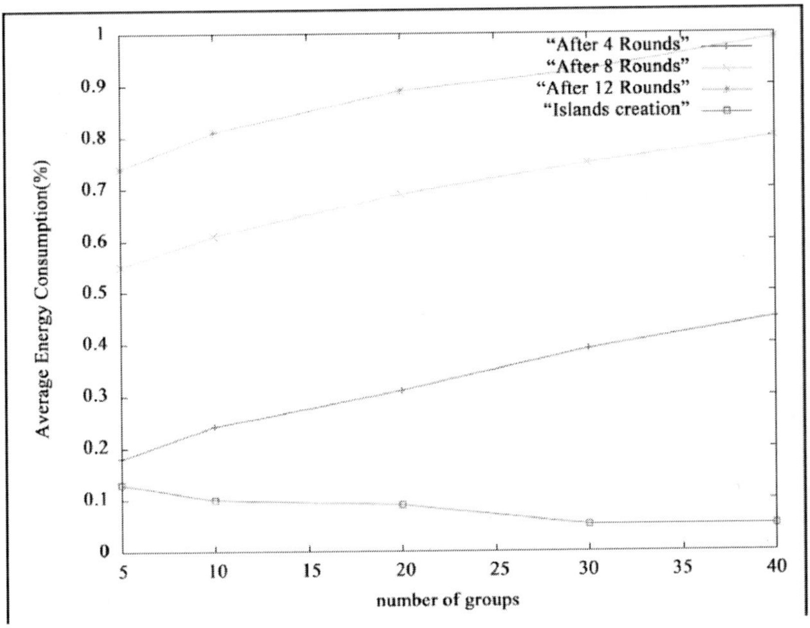

Figure 10. Average energy consumption.

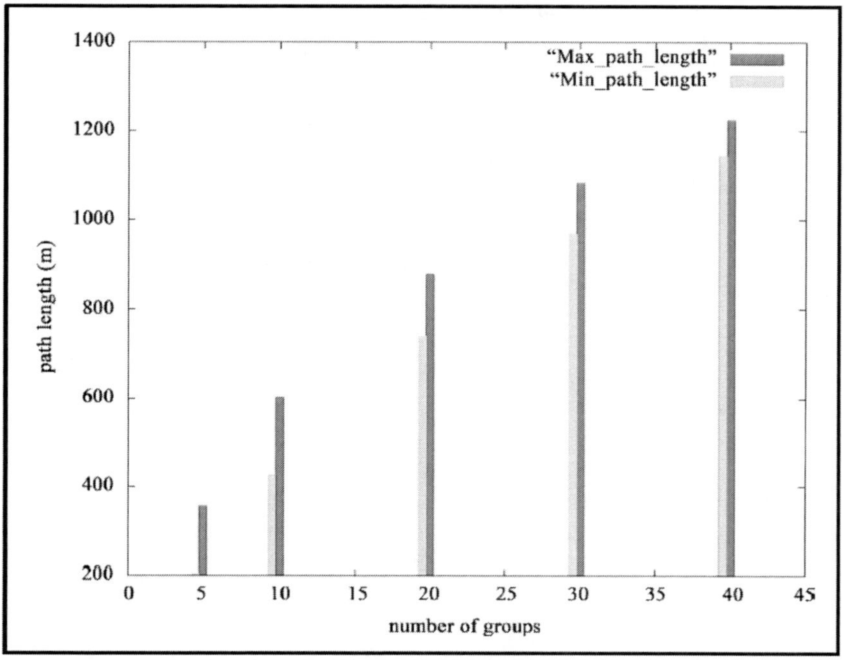

Figure 11. Maximal and minimal path length.

Figure 11 depicts the maximal and minimal path length according to the number of groups. This result has an effect on the data delivery delay from sensed nodes to the base station. The path length increases with the number of islands. However the gap between minimal and maximal path length decrease when the number of islands increases. In fact, regrouping more nodes in the same island means that the distance between nodes increase, so, the island master selection will affect considerably the path length.

6. CONCLUSION

The aforementioned approaches have incorporated context awareness for gathering data from an agricultural parcel. Uniform and dynamic parameters are defined. Depending on these parameters, data gathering strategy is defined with optimum energy preservation. Our approaches have been evaluated based on some simulations scenarios but the feasibility of the system and their productivity cannot be proved without real system tests. In order to prove their feasibility, our future work consists of modeling the system with an adequate analytical approach.

REFERENCES

1. Martinelli, M., Ioriatti, L., Federico, V., Benedetti, M. and Massa, A. (2009) A WSN Based Solution for Precision Farm Purposes. IGARSS, 469-472.

2. Zhang, W., Kantor, G. and Singh, S. (2004) Integrated Wireless Sensor/Actuator Networks in an Agricultural Application. Proceedings of the 2nd Academic International Conference on Embedded Networked Sensor Systems (Sensys 04), Acm Press, 317-318.

3. Anurag, D., Roy, S. and Bandopadhyay, S. (2008) Agro-Sense: Precision Agriculture Using Sensor-Based Wireless Mesh Networks. First ITU-T Kaleidioscope Academic Conference.

4. Burrell, J., Brooke, T. and Beckwith, R. (2004) Vineyard Computing: Sensor Networks in Agricultural Production. Sensor and Actuator Networks, 3, 38-45.

5. Lpez Riquelme, J.A., Soto, F., Suardaz, J., Snchez, P., Iborra, A. and Vera, J.A. (2009) Wireless Sensor Networks for Precision Horticulture in Southern Spain. Computers and Electronics in Agriculture, 68, 25-35.

6. Garcia-Sanchez, A.-J., Garcia-Sanchez, F. and Garcia-Haro, J. (2011) Wireless Sensor Network Deployment for Integrating Video-Surveillance and Datamonitoring in Precision Agriculture over Distributed Crops. Computers and Electronics in Agriculture, 75, 288-303.

7. Liu, Y., Zhang, J. and Richards, M. (2009) Towards Continuous Surveillance of Fruit Flies Using Sensor Networks and Machine Vision. Proceedings of 5th International Conference on Wireless Communications, Networking and Mobile Computing, 1-5.

8. Liu, H., Meng, Z. and Wang, M. (2009) A Wireless Sensor Network for Cropland Environmental Monitoring. Proceedings of the IEEE International Conference on Networks Security, Wireless Communications and Trusted Computing (NSWCTC 09), Wuhan, 25-26 April 2009.

9. Goumopoulos, C., Kameas, A. and O'Flynn, B. (2007) Proactive Agriculture: An Integrated Framework for Developing Distributed Hybrid Systems. Proceedings of the 4th International Conference on UIC, 214-224.

10. Dursun, M. and Ozden, S. (2011) A Wireless Application of Drip Irrigation Automation Supported by Soil Moisture Sensors. Scientific Research and Essays, 6, 1573-1582.

11. Ayday, C. and Safak, S. (2009) Application of Wireless Sensor Networks with GIS on the Soil Moisture Distribution Mapping. Proceedings of the GIS Ostrava Symbosium.

12. Cao, X., Chena, J., Zhang, Y. and Sun, Y. (2008) Development of an Integrated Wireless Sensor Network Micro-Environmental Monitoring System. ISA Transactions, 47, 247-255.

13. Brinis, N., Azouz Saidane, L. and Minet, P. (2012) EDGM: Energy Efficient Data Gathering with Data Mules in Wireless Sensor Networks. 6th International Conference on Sensor Technologies and Applications, SENSORCOMM 12, Rome, 19-24 August 2012.

14. Brinis, N., Saidane, L.A. and Minet, P. Adaptative Delay Aware Data Gathering Strategy for Wireless Sensor Networks. ICAIT12.

15. Commuri, S. and Watfa, M.K. (2006) Coverage Strategies in Wireless Sensor Networks. International Journal of Distributed Sensor Networks, 2, 333-353.

16. Ingole, P., Pawar, J. and Vhatkar, H. (2013) Context Aware, Mobile Phone Based Sink Model of Wireless Sensor Network for Indian Agriculture. Advances in Computing, Communication and Control Communications in Computer and Information Sciences, 361, 301-308.

6

DESIGN AND FIELD TEST OF A WSN PLATFORM PROTOTYPE FOR LONG-TERM ENVIRONMENTAL MONITORING

Mihai T. Lazarescu

Politecnico di Torino, Dipartimento di Elettronica e Telecomunicazioni, Corso Duca degli Abruzzi 24, I-10129 Torino (TO), Italy

ABSTRACT

Long-term wildfire monitoring using distributed *in situ* temperature sensors is an accurate, yet demanding environmental monitoring application, which requires long-life, low-maintenance, low-cost sensors and a simple, fast, error-proof deployment procedure. We present in this paper the most important design considerations and optimizations of all elements of a low-cost WSN platform prototype for long-term, low-maintenance pervasive wildfire monitoring, its preparation for a nearly three-month field test, the analysis of the causes of failure during the test and the lessons learned for platform improvement. The main components of the total cost of the platform (nodes, deployment and maintenance) are carefully analyzed and optimized for this application. The gateways are designed to operate with resources that are generally used for sensor nodes, while the requirements and cost of the sensor nodes are significantly lower. We define and test in simulation and in the field experiment a simple, but effective communication protocol for this application. It helps to lower the cost of the nodes and field deployment procedure, while extending the theoretical lifetime of the sensor nodes to over 16 years on a single 1 Ah lithium battery.

KEYWORDS

WSN; IoT; pervasive environmental monitoring; long-term operation; low maintenance; low-cost WSN nodes; low-cost WSN deployment; protocol design; low power

1. INTRODUCTION

Since the Internet of Things (IoT) paradigm was first coined more than a decade ago [1], wireless sensor networks (WSNs) were included among its enabling technologies as flexible, low-cost systems suitable to build dense and cost-effective monitoring infrastructures [2]. WSNs have relatively low material and deployment costs, since they do not need wiring, can self-configure and can operate for extended periods of time with very low, if any, scheduled maintenance. Sensor networks are also considered among the best techniques for environmental data collection for a variety of open-space applications, either close to human daily activities or remote, in the wild.

WSNs are typically made of multiple nodes, which are small and portable embedded computing systems interfaced with specialized transducers and radios for short-range communication. They are capable of autonomously monitoring, processing and transmitting various parameters at diverse locations for extended periods of time, using very limited energy and often without any maintenance in their lifetime.

More than a decade after the WSN vision, the application domain experts expect the technology to solve application demands inexpensively and with low effort. However, the multidisciplinary challenges raised by WSN optimization to multiple application needs have been constantly acknowledged over the past decade, especially for outdoor applications [3–5]. Most networks remain relatively small in size and/or are deployed for short periods of time, and node programming often requires experienced software engineers [6]. High cost and perceived low reliability are still important barriers [7], out of which maintenance cost may significantly increase in the case of remote deployments or in locations that are difficult to access.

As for most maturing technologies, WSNs reached the stage where standards consolidate most existing achievements. IEEE 802.15.4 and

6LoWPAN set the groundwork for applications with IPv6 support. Software-defined radio efforts build on these, aiming at more robust and versatile communication architectures through antenna-, frequency-, modulation- and data-rate diversity, on which services for diverse application spaces can be built.

At the same time, environmental monitoring has constantly grown in importance. Routinely collected data about the state and the evolution of the environment can help to detect and alert of hazardous events and to assess and optimize the long-term sustainability of, e.g., population growth, urban sprawl, invasive species, climate change and habitat destruction.

Nevertheless, the outdoor monitoring application domain is often challenging [8]. The quality of radio communication between nodes can be unpredictable, especially during atmospheric events (e.g., heavy rain, snow, frost, fog), due to plant and foliage growth or their movement due to wind, or due to the accumulation of dirt or insect nests on nodes. Repeated mechanical stress induced by large daily temperature variations may prematurely break the nodes, increasing the operating cost, since maintenance in open nature is usually difficult. Communication protocol scalability and the cost of nodes and field deployment may be stretched for applications that require thousands or tens of thousands of nodes for pervasive monitoring.

WSN platforms typically rely on existing hardware and software components to support a broad range of applications, often using some cross-layer optimizations. However, applications with stringent requirements, such as low cost, a large number of nodes, long service time, fast field deployment and reduced maintenance, can motivate custom design and optimization of the entire WSN platform.

In this paper, we focus on the needs of a demanding application, pervasive wildfire monitoring using *in situ* distributed temperature sensors. Early detection of forest wildfires is very important for fire fighting success and to effectively limit damage and cost. We comparatively analyze existing detection and alert solutions and define the design objectives of a WSN platform that can overcome most of their weaknesses. Then, we present the main design phases, the effort spent, the major pitfalls building the first prototype and the lessons learned from its field test. We also show how the resulting WSN platform can support a large class of environmental monitoring applications, increasing its value and reducing its return on investment.

Platform specifications were defined in collaboration with an industrial beneficiary and respond to pragmatic and stringent requirements for pervasive wildfire detection and alerting. For this purpose, the company consulted several actors in areas subject to wildfires, such as firefighters, state forestry corps, civil protection and forest rangers.

The main contributions of the paper can be summarized as follows:

1) A methodical holistic analysis of the effects of the application requirements and of the available design options for the platform hardware, software, communication protocols and field deployment procedures on the overall platform performance, the cost and the reliability of its nodes, of the platform commissioning and exploitation. This method can be applied for the analysis of other WSN platforms in the context of different application requirements.

2) Design criteria, optimization and field validation of cost-effective sensor nodes and gateways that use much lower resources and energy than the typical nodes in the literature based on off-the-shelf components.

3) Analysis, definition and field test of a cost-effective and reliable field deployment procedure for the network nodes that minimizes human-introduced errors.

4) Design, optimization and field test of a simple, yet effective field communication protocol that matches application requirements well, significantly contributes to the reduction of node energy consumption and cost and supports an effective field deployment procedure for the nodes. It can also scale up to support networks with a very large number of nodes.

5) Analysis and selection of the hardware and software features to implement on node prototypes designed for laboratory testing in order to adapt them, in a short time and with reduced effort, for a relatively long, unattended field experiment with two equally important purposes: to check and demonstrate the autonomous operation of the platform and to acquire sufficient visibility of node and system behavior that can be used to improve the design of the platform.

The rest of the paper is organized as follows. Section 2 reviews related published works and defines the specifications of the system through comparative analysis of the strengths and weaknesses of existing methods. Section 3 outlines the design criteria, which drive the platform design decisions. Section 4 explores the most important design decisions made for the implementation of the WSN platform components. Section 5 presents how the prototype platform was prepared for a field test, on short notice. Section 6 presents the unfolding of the field test, the major issues during its various phases and the most important lessons learned to improve the WSN platform. Section 7 presents a summary of the future work and concludes the paper.

2. SYSTEM DEFINITION

To better define the specifications of the proposed wildfire monitoring method, we will briefly review several of the most representative existing techniques with a focus on cost, detection reliability and effectiveness for early alerting.

2.1. Wildfire Detection Methods

Wildfires are quasi-natural hazards likely to occur under specific climatic, weather and vegetation conditions [9]. A small fraction ignite from natural causes that include lightning, volcanic eruption, sparks from rockfalls and spontaneous combustion, but most ignite from human activities, such as arson, discarded cigarettes, sparks from equipment, power line arcs and controlled field burning [10–12].

While most common wildfire causes may vary throughout the world [13,14], early detection is always very important, because fires may spread fast and escape firefighters' control, potentially causing significant economic damage, threatening human lives, homes and resources and devastating wildlife, habitats and ecosystems [15].

Wildfire frequency and devastation make the continuous monitoring of the areas at risk one of the most well-known environmental monitoring activities. Although many detection techniques were proposed, cost-effective and reliable early wildfire alerting is still difficult to achieve.

2.1.1. Ground-Based Remote Detection Methods

Traditionally, wildfire monitoring is based on continuous human surveillance from selected vantage points in the field attempting to recognize smoke patterns during the day or the light from fire flames during the night using visible light [16,17]. The authors argue in [18] that observers' reports are often insufficient to define an effective firefighting strategy and that the field video surveillance monitored in the operational center can provide richer, actionable information and increase the surveyed area by the operator. However, since remote field monitoring increasingly subjects the operators to fatigue, boredom and loss of concentration, several semiautomated wildfire recognition systems are proposed [19–34]. These systems reduce the effort of the analysts, which basically need to validate the positives triggered by the automated detection methods [22].

A class of detection methods look for combustion-generated heat using infrared (IR) imaging. The equipment is rather expensive, has reduced spatial resolution and is prone to false positives [21,23]. Accuracy can be improved by automated cross-matching of IR images with visible or ultraviolet (UV) ones or with meteorological and geographical data using various processing and decision making techniques, but at a higher equipment and maintenance cost [24].

Optical spectrometry can detect combustion-specific products in smoke [25,26] by analyzing its sunlight absorption patterns. It has good reliability, but is limited to scans above the horizon (thus, for smoke that clears the tree line), and at night, it switches to the less effective flame detection [27]. Strong winds may lower or disperse the smoke, delaying detection. However, wind significantly increases the speed of fire spreading; thus, early alerting of firefighters becomes crucial for blaze containment.

Light detection and ranging systems (LiDAR) analyze the laser light backscattered by smoke [28,29]. High-power laser beams may be expensive to produce, although, recently, researchers proposed lower cost techniques [30], and they can also be a public health hazard. Its effectiveness may be limited by obstacles in the line-of-sight, fog or low altitude clouds, which are a likely occurrence in mountain climate.

Radio-acoustic sounding detection (RASS) can detect crown and surface wildfires from thermal maps built by inferring air temperature from shifts in the propagation time of acoustic waves [31]. The system can account for atmospheric events, like wind and some static echoes from vegetation or landscape, but it requires a clear line-of-sight right above the tree tops for proper operation. Moreover, its deployment and calibration can be complex.

Remote detection systems tend to generate false positives (especially those based on visible light image analysis) and need the support of human analysts. Those that are more accurate are also more expensive due to higher equipment and maintenance cost [35].

2.1.2. Overhead Remote Detection Methods

Satellite-based wildfire detection can reach low false alarm rates in near real time. However, geostationary satellites may have limited spatial resolution, while polar orbiting satellites have limited temporal resolution, which can be significant for rapidly-evolving wildfires [36,37]. The accuracy of the automated methods based on satellite monitoring can be further reduced by image acquisition problems and image navigation errors, which can compromise the monitoring of individual fire events that would otherwise be correctly identified by analysts [38].

Unmanned aerial vehicles (UAV) can efficiently monitor for wildfires over large areas [39,40]. Equipped with various fire detection techniques, they can survey the area continuously and cost effectively from a diversity of vantage points, much closer than the satellites. UAVs operation can be affected by local weather conditions (fog, strong wind, low altitude clouds), wind being a well-known risk factor during their takeoff and landing [41].

These methods also share most operating limitations of the ground-based remote detection techniques mentioned in Section 2.1.1, like unfavorable weather, obstructions of the direct line-of-sight, ambient light quality and incidence angle.

2.1.3. Pervasive *In Situ* Detection Methods

In situ pervasive wildfire detection has several advantages over remote detection [42]. First, the sensors are closer to the events (e.g., combustion); thus, their sensing is generally less susceptible to interference from external factors (e.g., wind, fog, obstructions). Second, *in situ* detection is implicitly redundant to some extent, since the higher sensor density makes it likely that more than one of them detects an event within a short time window. Nevertheless, natural occurring, non-fire-related phenomena (such as fog, dust or insect nests) may still trigger false alerts, even for pervasive methods, or the effectiveness of some detection methods may be limited at night or in low visibility conditions.

Moreover, it is important to note that the cost per unit of surface of *in situ* monitoring is usually higher than that of remote sensing, both for the initial deployment and for the maintenance [43]. Thus, reducing these costs can significantly extend the application range that can be economically addressed using pervasive methods. Their finer spatial and temporal resolution and the generally higher detection accuracy can complement the remote detection methods and can be justified for areas with higher risk of fire or where humans or goods are more exposed, such as along roads or trails that cross forests or at the interface between human habitat or resources and forests.

WSNs are among the best-suited systems for pervasive *in situ* long-term environmental monitoring, even in wild and harsh conditions [44–50], although other techniques have been reported, as well, e.g., based on optical fiber [31]. WSN nodes typically include the transducers and processing power to autonomously detect specific fire effects within their range, such as light, smoke, heat and noise [51–55]. The nodes are deployed within the area to monitor using application-specific criteria [56] and periodically check for fire signatures using their on-board sensing and processing capabilities. The control center is notified of positive matches by using radio messages sent either directly to gateways or forwarded through a mesh network made by their neighboring nodes.

Most fire detection techniques used by pervasive WSNs are shared with the remote methods (see Sections 2.1.1 and 2.1.2) with various degrees of detection confidence, power profile and cost [49,57]. Low-power techniques are preferred, since the limited sensor node energy is an important selection factor, energy being a direct and indirect contributor to WSN solution cost (we will analyze this in detail in Section 3.1.1).

Techniques based on low-power image processing [58] or radio-acoustic detection of the sound of burning fire [31] performed on board the sensor nodes are reported, but they generally require higher processing energy. Detection of infrared [59] or ultraviolet radiation

[60] that are correlated with fires have a lower energy consumption profile and can be used alone or in conjunction with smoke detection for increased accuracy. Smoke particle detection can be combined with temperature sensing [53] (either as absolute value or variation speed). The detection of temperature patterns can be used alone (and this is arguably one of the most cost- and energy-effective techniques) [61], or correlated with humidity readings [44], or with humidity, wind and rainfall data that are periodically retrieved from the application server [48]. In general, techniques based on data fusion from multiple transducer types can increase detection reliability [35,62,63]. Reliability can be increased also using adaptive methods on sensor nodes [64,65] or on sensor and gateway nodes, the latter benefiting from data aggregation from multiple sensor nodes [66].

Other methods are based on the detection of abnormal levels of combustion gases, standalone or combined with temperature detection [67]. However, gas transducers may require higher power and periodic recalibration. WSN detection can also be combined with UAV monitoring [50] to increase the detection accuracy and to collect the necessary data to define and coordinate the actions of the firefighters. While most WSN deployments are designed for long-term monitoring, they can be deployed in the affected areas for the short term to continuously collect data to help support firefighters' operations in the field [68].

Meeting the stringent requirements for effective long-term outdoor environmental monitoring is generally challenging for WSN technologies [69–71]. Nodes need to reliably operate for very long times without or with minimal maintenance in variable and sometimes extreme climatic conditions. Outdoor applications demanding a high number of nodes put additional emphasis on node cost and reliability. Node weight, size and ruggedness become also important for deployments in fields that are difficult to reach, e.g., when nodes are transported in backpacks [8].

2.2. System Specifications

The analysis in Section 2.1 shows that *in situ* pervasive sensing is among the fastest and most reliable monitoring methods. A low false alert rate due to higher detection reliability can reduce the need for human analysis and the effort of the field personnel, thus reducing the exploitation cost. Earlier fire detection can help the firefighters to contain the risks and damages, increasing safety and reducing costs.

WSN technology looks well suited for distributed sensing and reporting, but to be economically viable, it needs to scale well for large deployments and to have reduced initial and operation cost: low sensor node cost, reliable operation for extended periods and to require very few or no maintenance at all.

Transducers on board the WSN nodes typically use energy sparingly to preserve their reserves and life time. Among the effects of the fire, heat is always present, and high temperatures are among its most devastating effects. The temperature of the surrounding air can be energy- and cost-effectively measured (e.g., using inexpensive negative temperature coefficient (NTC)transducers), and the WSN nodes can periodically sample it looking for abnormal temperature patterns that are highly correlated with convection heating from fires. For example, the nodes can alert when the absolute temperature value or its rate of increase rises above specific thresholds [53]. The thresholds should be set high enough to prevent false alerts due to naturally occurring temperature patterns and as low as possible to improve detector sensitivity and timely alerting.

As discussed in Section 2.1.3, fire detection accuracy can increase by combining measurements from different types of transducers, each sensitive to different effects of the fire. For instance, abnormal air temperatures can be correlated with the levels of infrared radiation on wavelengths closely related to the presence of fire, for the same monitored area.

Transducers of different types can be mounted on the same sensor node or on different, specialized nodes. A node with multiple transducers can correlate the measurements on board, reducing network traffic. However, multiple transducers also increase the node cost, energy consumption and resource requirements. They may also add deployment constraints, increasing its complexity and cost.

Specializing the nodes by transducer type allows one to differentiate their deployment requirements to optimize their operation. For instance, nodes that sense infrared radiation can be attached to the underside of the lower branches of the trees, facing down, at heights of a few meters above the ground, while the temperature sensors can be attached on tree trunks, at 0.5–1 m from the ground, to have good thermal contact with the hot air rising from nearby fires. The deployment density may also differ by node type. Field data from nodes can be correlated upstream, e.g., on the gateway or server for a typical tiered structure, like the one shown in Figure 1.

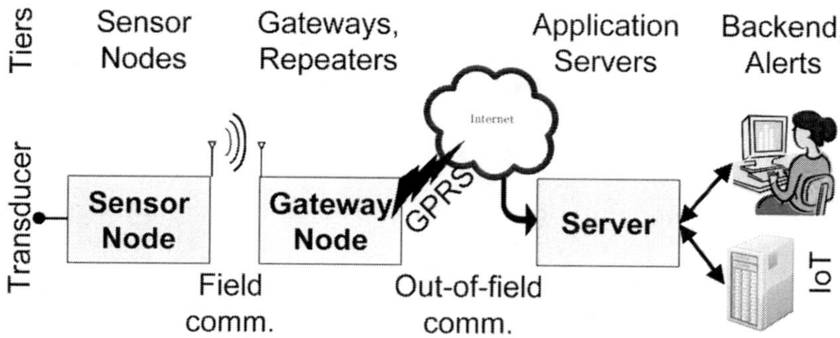

Figure 1. Typical tiered structure of WSN platforms.

Application-specific trade-offs are usually made between the required detection accuracy and the implementation cost. The industrial partner initially considered fire detection solutions based on low-cost sensors, which consume no energy in standby mode and that are activated by fire (e.g., heat-activated power supplies [72] or conventional small batteries that supply the node through a bimetallic switch only when heated above a threshold [73]). However, the quality of service (QoS) cannot be monitored for these solutions, thus they cannot be considered reliable in the general case.

Consequently, the specifications evolved towards more typical WSN platforms that use active sensor nodes and a tiered structure as shown in Figure 1. Sensor nodes are fit with transducers to sense the environment properties of interest and can communicate with the gateway (or sink) nodes using short-range in-field radio communications, either directly or through multihop message forwarding. The gateways may further process and aggregate the data and can communicate with the application server over long-range data channels. The server has many functions, such as reliably storing the field data, providing suitable user interfaces for observation and analysis, data post-processing, configurable alerts and interfacing with other IoT applications.

From the comparative analysis in Section 2.1, the requirements of a WSN platform well suited for reliable and cost-effective long-term pervasive environmental monitoring can be defined as follows:

- low-cost, small-sized nodes (sensors and gateways) that can process the transducer data on board, perform self-tests and recover from errors (preserving field data if possible);
- receive software updates over the air with minimum service disruption and have one or more long-range communication channels (at least the gateway nodes);
- the field communication protocol should efficiently support:
    - network sizes from a few sparse nodes up to a very large number of nodes;
    - low data traffic per node, in small packets;
    - resource-constrained reliable operation, with small memory and energy overheads;
    - reliable operation in case of node failures, especially if several nodes are destroyed by fire;
- transmit events instead of raw data to reduce communication energy by processing the data close to the source (on sensor nodes and gateways);
- fast and reliable field deployment procedure for network nodes;
- adequate gateway hardware and software resources and interfaces for application-specific extensions (e.g., local transducers, data storage, communication channels, other peripherals);
- node ruggedization to withstand long-term exposure to environmental conditions;
- long maintenance-free service (the beneficiary requirements were 5–10 years for the sensor nodes);
- high availability of service for nodes and servers;
- multimodal access to server data for human interfaces and automated processing;

- programmable multichannel alerts on field events;
- automatic detection and reporting of platform faults;
- extensible server architecture for integration in IoT applications.

These specifications would also allow the reuse of the platform for several event-based long-term applications besides wildfire monitoring, such as:

- water level of lakes, streams, sewage. Sensor nodes can detect and alert about extreme levels (high or low) and periodically report level statistics;
- gas concentration in air for laboratories, deposits, ship holds, cities. Sensor nodes can detect potentially harmful concentrations of one or more gases;
- soil properties for precision agriculture;
- static or dynamic parameters for structures, e.g., the inclination of dams;
- position changes, e.g., land slides;
- lighting level, e.g., for intrusion detection in dark spaces, like cable channels;
- infrared radiation from combustion (fire) or human or animal presence, for security or habitat monitoring. On-node analysis using thresholds and speed of variation can screen alerts from natural-occurring patterns.

Platform reuse can improve the economic sustainability of its development and maintenance. This is important, since cost can be a major factor that prevents commercial adoption of *in situ* pervasive monitoring solutions.

The overall cost of the platform has several components that we consider in detail during platform design in Section 3:

- Node cost, which is roughly proportional to the monitored area and can significantly weigh on the total cost. Generally, a WSN has fewer gateways than sensor nodes; thus, platform design needs to carefully optimize at least the cost of the latter.

- Exploitation cost basically made up of the cost of the analysts to validate the alerts and the cost of node and network maintenance. The rate of false alerts can be kept low through several techniques: performing the sensing close to the event source, using selective sensing methods less susceptible to interference from unrelated events and using data fusion from multiple sensor types [74]. Maintenance may be needed to periodically clean the nodes to restore their nominal sensing or energy harvesting levels or to replace exhausted batteries. This can be reduced by using maintenance-free transducers and low-energy long-life sensor nodes. Nodes can fail for various reasons, including mechanical stress due to daily temperature variations or extreme climatic conditions. Adequate node hardware design and ruggedization should minimize the effects of long-term environmental exposure.

- Deployment cost, which can significantly increase when the monitored area is difficult to access or if deployment errors are discovered later, during network exploitation, requiring additional field expeditions to correct them. Deployment cost depends also on other factors, such as the complexity and time needed for the deployment of the network nodes, the level of specific knowledge expected from the deployment personnel and node size and weight (if they are manually transported in field).

Most platform costs can be reduced through optimizations of the design of its components with a holistic view of the full system, as will be shown in Section 3.

3. PLATFORM DESIGN CRITERIA

We will present in the following the most important node design considerations and decisions based on the specifications in Section 2.2.

3.1. Sensor Node Design Criteria

WSN platforms for pervasive *in situ* monitoring may need to deploy many sensor nodes, which can affect several cost components of the platform. In the following, we will analyze the design options for the major contributors to sensor node cost, as well as their contributions to the various components of the platform cost that were presented in Section 2.2.

3.1.1. Sensor Node Energy Supply

Energy supply can be an important cost of the sensor nodes. Its role is to provide energy for node operation between maintenance (e.g., for battery replacement or node cleaning), with the continuity necessary for the expected node quality of service (QoS).

An important decision is whether to use a regenerative supply. Energy harvesting can be effective when ambient energy is available, such as vibration, fluid flow, temperature differences and light [75]. Although their cost and QoS are difficult to estimate in the general case, combined energy harvesting can increase the availability [76]. These supplies may add to node or platform cost, for instance by increasing the physical dimensions of the node, by additional casing requirements (e.g., a waterproof transparent window for light harvesting), by additional restrictions for field placement for effective ambient energy harvesting or by increasing the complexity of the deployment procedure.

Primary batteries, on the other hand, store all of the energy needed for the entire node lifespan or the period between scheduled maintenance. The minimum battery capacity can be obtained by multiplying the node average energy consumption and the expected lifespan. Battery cost is roughly proportional to its capacity, but the scheduled maintenance for replacement of depleted batteries can be a major contributor to the overall solution cost.

3.1.2. Sensor Node Communication Protocol and Radio Device

Radio communication is widely recognized as one of the major energy drains on sensor nodes. Thus, it influences the selection of the energy supply (with the criteria presented in Section 3.1.1) and some other node costs that we will analyze in the following.

The largest part of the communication energy is typically spent by the receiver while listening for incoming data packets. For this reason, most communication protocols operate the receiver with very low duty cycles (typically 1%–3% for random MACs [77,78] and much less for synchronous MACs [79]). Even more, the receiver can be eliminated altogether for random access unidirectional communication in star topologies.

Multihop message forwarding typically uses less energy to cover the same distance and may be less affected by node faults or changing propagation conditions. However, mesh networks require more microcontroller resources and involve regular packet exchanges to maintain network connectivity. These may increase both node energy consumption and cost. Moreover, message routing in large mesh networks may be affected by dynamic effects that are hard to detect, such as bottlenecks or instabilities [8,80,81], which may increase deployment complexity and cost. Synchronous protocols also need an accurate timer constantly running on the nodes, which further increases their cost and energy consumption.

Transmit-only nodes in star topologies can use very small protocol stacks and have much more relaxed requirements for on-node timing accuracy. Thus, they need less microcontroller resources and energy, but they cannot create mesh topologies to improve robustness and reduce transmit power. Furthermore, these nodes cannot actively avoid message conflicts or receive confirmations, queries, configuration changes or in-field updates for node software.

3.1.3. Sensor Node Functions

Besides the main function of environment sensing and processing, sensor nodes can implement functions that improve the overall system reliability and cost.

Periodic node self-test and error recovery functions increase the reliability and QoS of the monitoring platform and reduce maintenance requirements and cost. Nodes that aggregate peer messages before forwarding should implement suitable redundancy policies to prevent data loss if they reset themselves in order to recover from errors or if they fail.

Normally, the sensor nodes do not have user interface elements, such as buttons, LEDs and displays, which can increase the bill for materials, the ruggedization cost and the energy consumption. Thus, for interactive operations, like checking the quality of an installation position in the application field or retrieving their ID, the nodes should provide the operator simple, yet effective means to selectively address and communicate with them while in close physical vicinity. Moreover, specific node self-testing and reporting functions should be used at the end of the production line to lower production cost.

However, additional functions increase the complexity of the node program, thus the probability of defects [82], which can adversely impact the overall system reliability and increase the maintenance cost. Programming the nodes using high-level, clean, application-optimized code and component reuse may require less resources and energy, thus reducing the cost and improving the reliability of the node.

3.1.4. Sensor Node Hardware Components

In the following, we examine some criteria that guide the sensor node hardware selection.

Smaller bills for materials generally increase the production and operation reliability and reduce costs.

Transducer selection closely depends on the detection method. For instance, fire can be sensed through smoke or heat, and the latter can be sensed by temperature (e.g., using an NTC) or by radiation (e.g., using passive infrared transducers). Low-cost and power-efficient transducers, which can operate on very low duty cycles and with less energy, can significantly reduce node and maintenance cost.

Radio devices with a baseband modem may reduce the software complexity and energy consumption.

The microcontroller resources should fit the application needs closely, since oversizing it may increase the cost and energy consumption on most power levels.

3.2. Gateway Design Criteria

In typical WSN deployments, there are significantly fewer gateways than sensor nodes. Hence, the node cost of the gateways contributes less than the cost of the sensor nodes to the overall cost of the WSN platform. However, the features of the gateways can indirectly influence the cost of the platform in several ways, as we will show in the following.

3.2.1. Gateway Communication

The gateways typically have two types of communication capabilities: long range and short range. The long-range communications are used to connect with the application server either directly or through the Internet. The short-range communications are mainly used to connect with the field nodes, either sensor nodes or peer gateways.

Long-range communication can use several technologies. A cellular data network is commonly used in outdoor fields that are covered by mobile data services [83]. Alternatively, dedicated communication links on either a private or public radio spectrum can be used, e.g., the 868-MHz short range devices (SRD) band allows ranges up to 40 km [84], and long-range Wi-Fi links can exceed 100 km [85]. Satellite links may also be used [86], especially for remote, isolated fields.

The constraints for the short-range in-field communications are less stringent for the gateways than for the sensor nodes (which were discussed in Section 3.1.2). The gateways have more energy reserves and computing resources, thus they can support most communication protocols: from simple low power listening (LPL) for random access MACs in star topologies and up to complex routing and synchronization schemes.

Thus, the gateways can help reduce the overall cost of the WSN application by supporting the short-range communication protocol that optimizes the cost of the sensor nodes, which is a significant component of the overall cost of the solution.

3.2.2. Gateway Energy Supply

Long-distance communication of the gateways typically requires much more energy than the short-range in-field one of the sensor nodes. Gateways also have more hardware resources, such as larger program and data storage capacity, more powerful microcontrollers or microprocessors and accurate, always-running timers. All of these combined require significantly more energy than the sensor nodes.

Thus, the gateways generally have higher energy reserves and may also have ambient energy harvesting devices to increase their lifespan between scheduled maintenance (which may be needed for, e.g., the replacement of the (rechargeable) battery or for the cleaning of the energy harvesting devices). Even if the gateways may require scheduled maintenance more often than the sensor nodes, its cost may be acceptable: a WSN deployment for pervasive environmental monitoring typically has fewer gateways than sensor nodes, and the gateways are generally deployed in locations that are more accessible than the sensor nodes (see Section 2.2), which further reduces the cost of gateway maintenance.

Given that gateway energy reserves are far larger than those of the sensor nodes and that gateway maintenance is less expensive than the maintenance of the sensor nodes, it can be cost-effective at the network-level to bias the energy consumption towards the gateways if this helps with reducing the cost of the sensor nodes and their scheduled maintenance requirements.

3.2.3. Gateway Functions

Besides the main function to collect and forward to the application server the field data from the sensor nodes (and from the on-board sensors, if any), the gateways can implement other functions that improve the overall system reliability and cost.

The gateways should prevent the loss of the aggregated field data in case of non-fatal hardware faults or of self-resets due to software defects. Moreover, in case of repeated or fatal software errors, the gateways should fall back to a safe operation mode, which allows remote debugging or recovery to reduce service interruptions and the cost of in-field maintenance.

During field deployment, the gateways should assess the suitability of the installation position in terms of network reachability and link quality with the other nodes and the server, as well as the effectiveness of the energy harvesting (if available).

3.3. Server Design Criteria

The main purpose of the application server is to receive field data and store them reliably for later access upon request. It bridges the low-power communication segments (that operate under strict latency-energy trade-offs) with the fast and ubiquitous data access needed by human interfaces or IoT applications. As shown in Figure 1, the server provides interfaces for field nodes, operators and supervisors of the field, alert channels and for external systems.

For a high QoS, the server should implement high-availability techniques, e.g., a secondary server should be kept in sync and automatically take over the service if the main server fails.

The protocol between the nodes and the server should be simple to reduce the resource requirements for the nodes and, as much as possible, stateless, because the long-range connections at the fringe of mobile data coverage can be often unstable and the overhead for resuming an interrupted transfer should be minimized. For the same reason, the protocol should provide acknowledgments for data reception, and these should be sent only when the data are successfully replicated on the secondary server to avoid field data loss due to single points of failure.

Gateways that need to enter the fail-safe operation mode may use a communication protocol with the server that is optimized for the operations executed in this mode (e.g., software update). The server should implement this protocol and the associated operations, as well as promptly notify the operators when a gateway connects using this channel, to take corrective actions.

4. PLATFORM DESIGN

Platform design aims at a cost-effective implementation that closely follows the specifications in Section 2 and the design criteria in Section 3.

Considering the multi-objective optimization needed to satisfy sensor node specifications and design criteria, we decided for a custom design and development of communication protocols (both for in-field communications and with the server) and node hardware and software.

In the following, we describe the design of the prototype platform, the field test preparation and unfolding, the lessons learned from the test and how we can use them for improvements.

4.1. Communication Protocol Design

The communication protocol can affect the performance of a WSN platform in several important ways. First, the protocol is mostly responsible for propagating the data across the network according to application requirements. Consequently, it controls the most part of the radio activity of the nodes, which is generally recognized as a major energy consumer. Thus, the protocol can significantly influence the selection of energy supply type and size, which relate closely to its cost and to the scheduled node maintenance. Moreover, the characteristics of the communication protocol may also affect the complexity and quality of the field deployment of the network, thus its duration, cost and the need for subsequent corrective network maintenance [6].

Most WSN deployments tend to use simple communication protocols that are enough to satisfy the requirements of the applications [87]. However, very few appear to consider during network design the effects of the protocol on the complexity, quality, maintenance requirements and cost of network deployment [43]. Deployment and deployment-related maintenance can add up to significant costs for pervasive environmental monitoring applications.

The analysis in Section 3.1.2 presents some of the most significant trade-offs to consider when selecting the protocol for a specific application. For instance, using a radio receiver can increase the cost of the sensor node and of the application by adding to the cost of components, of the microcontroller (may require more resources), increases node energy consumption, thus adding to the cost of energy supply and/or the scheduled maintenance to change node batteries. However, a receiver also allows a better management of the communication channel, can increase the reliability of message delivery, can reduce the transmission power per node through multihop message forwarding and can support bidirectional network communication that allows one to remotely query or reconfigure the nodes in the field.

To define the specifications of the communication protocol, we started from the analysis of application requirements. The communication needs of sensor nodes by state can be summarized as follows:

- normal state: the node periodically communicates its health status and, if possible, some background environmental measurements;
- fault state: the node communicates the nature of the fault (if the fault does not affect the radio); alert state: the node detected a fire and reports the event in the shortest time, with the highest priority.

If the nodes periodically communicate their health status and some background environmental measurements, remote node interrogation becomes less important or even impractical for a large number of nodes, which are typical for pervasive monitoring methods (from hundreds to tens of thousands of nodes per application). Thus, for such applications, a bidirectional communication may be less important than the optimization of node cost.

Multihop message propagation through the network using a mesh topology can reduce the transmit power of the nodes, which may lead to lower energy consumption at the node level. However, two other factors should be considered. First, operating a radio receiver, even with very low duty cycles, can still significantly raise the energy consumption of the node. Second, lower radio frequencies, which propagate better in forest environments [88,89], have also a considerably longer communication range for the same transmit power, thus making the use of a star topology more practical. Moreover, a network with a star topology can be functional, even with sensor nodes with no receive capabilities, which can lead to multiple cost reductions, as we have shown.

Thus, on the one hand, we decided to use the 433-MHz UHF industrial, scientific and medical (ISM)band for its better propagation in forests. On the other hand, we analyzed if a communication protocol based on unscheduled channel access can satisfy the requirements of the application.

Communication protocols with unscheduled channel access are well known in the literature [77]. In the following, we will analyze if we can completely remove the receiver from the sensor nodes to achieve multiple cost reductions.

Without receive capabilities, the sensor nodes have no means to know if a message reached its destination or was lost. Thus, sufficient communication redundancy should be built into the protocol to keep message loss to acceptable levels for the application. It should be noted that the application is more tolerant to the potential loss of low-priority periodic health status messages than to high-priority and delay-sensitive fire alerts.

Higher transmission redundancy may increase the delivery probability of the messages, but it also increases the energy consumption of the node and the probability of collisions. Thus, protocol design should consider the effects of redundancy on message loss and on the delay of the detection of faulty nodes that are no longer able to transmit their status, for various network sizes.

For this purpose, we have set up a network simulation model in SystemC [90] as follows. The sensor nodes were modeled as transmitters that periodically send a health status (heartbeat) message. The transmission period is affected by a static error that is initialized randomly in the range +10–+30%, corresponding to the error range of the watchdog timer of the microcontroller that was used to control the timing on the sensor node.

The gateway node was modeled as a receiver implementing LPLto listen for incoming messages from the sensor nodes. The wakeup period for RF sampling was set slightly shorter than the length of a message, 0.3 s (including the preamble). Normally, the radio receivers can correctly decode the strongest of two overlapping messages if the amplitude difference exceeds the receiver signal-to-noise ratio, thus losing only one of the two. The same stands for messages that overlap only partially over the preamble. However, for the purpose of the simulation, we opted to pessimistically consider that two overlapping messages are both lost, regardless of the amplitude ratio or the extent of the collision.

Figure 2a shows the simulation results for a year of operation for fields with one gateway that receives heartbeats from 10–5000 nodes. The node heartbeat period varies from six minutes to one hour, and we count the number of times that a sensor node is declared missing over one year of operation for different time windows in which the gateway expects to receive at least one heartbeat from a node (to avoid cluttering Figure 2a, we have plotted only the results for three windows, of 45, 101 and 228 min).

We see that for a given heartbeat period and gateway window, the erroneous node missing events increase with the number of nodes. This is expected, since more nodes randomly sending heartbeat messages increase the probability of collisions, thus of message loss.

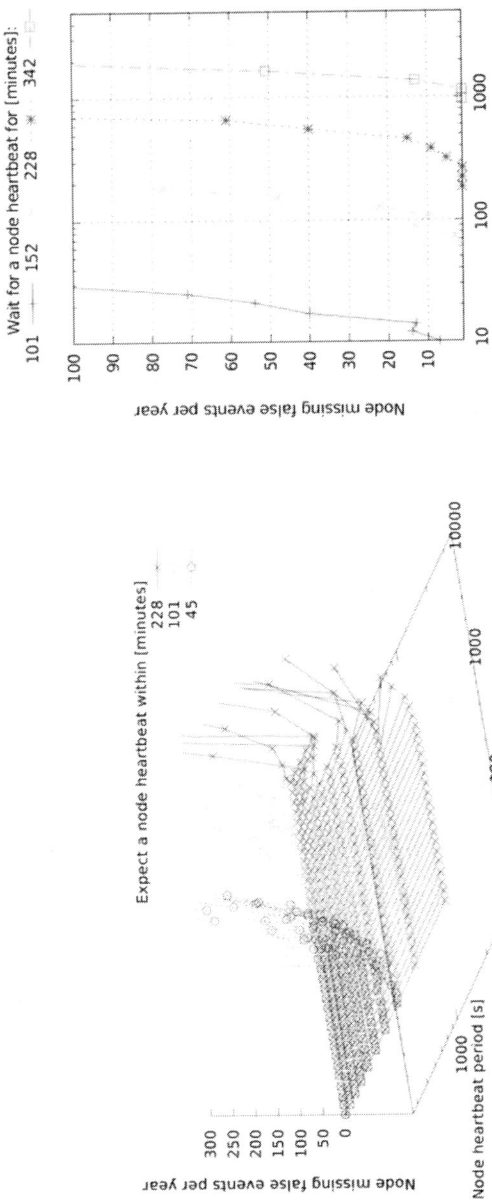

Figure 2. Simulation results of the field communication protocol for design space exploration. The rate of the missing node erroneous reportsincreases with the number of nodes and with the heartbeat period and lowers with the increase of the heartbeat listening window of the gateway (**a**); missing nodes can be detected in less than 6 h with a negligible rate of erroneous reports, for a 1 h heartbeat period (**b**).

We also note an increase of erroneous node missing events with the increase of the sensor node heartbeat period for a given number of sensor nodes in the field and a given gateway window. This result can be counterintuitive, since longer heartbeat periods mean less concurrent messages contending for the communication channel on average, thus lower collision probability. However, considering that the gateway window for receiving a heartbeat message from a sensor is constant, a longer heartbeat period means also that fewer heartbeats are sent within the window, which effectively decreases the protocol redundancy that is needed to overcome message losses typical for unscheduled MACs.

We used these simulation results to choose a compromise between the frequency of sensor node heartbeats (which contribute to node energy consumption), the delay after which the gateway detects missing nodes and the frequency of false node missing reports. As we will show in Section 4.2, one heartbeat per hour increases node energy consumption by about one third, which we consider acceptable for this application. Furthermore, in Figure 2b, we see that for a network traffic made of one heartbeat per hour per sensor node and a gateway window of slightly less than six hours, we achieve very low false positives for fields up to about 1000 nodes. This is realistically large, since we expect at most a few hundred nodes per gateway in practical deployments. Furthermore, a six-hour delay in reporting missing nodes is considered suitable for scheduling network maintenance operations.

Thus, the main characteristics of the communication protocol can be summarized as:

- under normal conditions, the sensor nodes send one heartbeat per hour;

- fire alert messages are sent with much higher frequency (every few seconds) to ensure their timely propagation, even at the cost of losing concurrent heartbeat messages. The communication channel capacity can support the alert traffic well, since, assuming no collisions, 1000 nodes sending one 0.3 s heartbeat message per hour each occupy at most 8.3% of the channel capacity;

- the gateways can detect a missing node in less than six hours.

This simple protocol satisfies the requirements of the application well. It does not need a receiver on the sensor nodes, which primarily lowers node cost and energy consumption. Moreover, without a receiver, the operation of the sensor nodes is defined only by their own programming. This means that the energy consumption for each node in normal conditions is predictable, and we calculate it in Section 4.2. In the alert state, the node will spend much more energy in order to propagate in a timely manner the alert messages to the gateways, which can significantly reduce its energy reserves. However, this is not an issue, since the heat of a fire will very likely destroy the node shortly after detection.

4.2. SENSOR NODE DESIGN

Early node prototypes were based on zero standby energy designs discussed in Section 2.2. Since their health status (QoS) cannot be observed, they were abandoned for a typical WSN node architecture: a microcontroller communicating with a radio device and powered by a primary cell, as shown in Figure 3.

The prototype was designed mainly to test the operation of the low-resource microcontroller and radio in a compact format, including the radio frequency (RF) performance, and to develop the node firmware, software and in-field communication protocol. Thus, the node printed circuit board (PCB)did not include the transducer, the power supply pads did not fit a standard battery size and it had just generic soldering pads for the antenna (all of these elements were still to be decided).

According to application specifications, the node prototype was primarily optimized for cost and size. In this regard, it is worth noting that it has significantly lower resources than a typical WSN node: the microcontroller is an ATtiny25 (2 KiB FLASH, 128 bytes SRAM, 128 bytes EEPROM, one analog to digital converter (ADC), one serial peripheral interface (SPI)); the radio for in-field communication is a TI CC1150 with a baseband modem (transmit-only for the 315-/433-/868-/915-MHz bands); and the external accurate resonator for the radio is a 27MHz quartz crystal. The microcontroller uses its internal oscillators (8MHz main clock and 128 kHz watchdog clock) to reduce the external components, the wake-up time and the energy consumption.

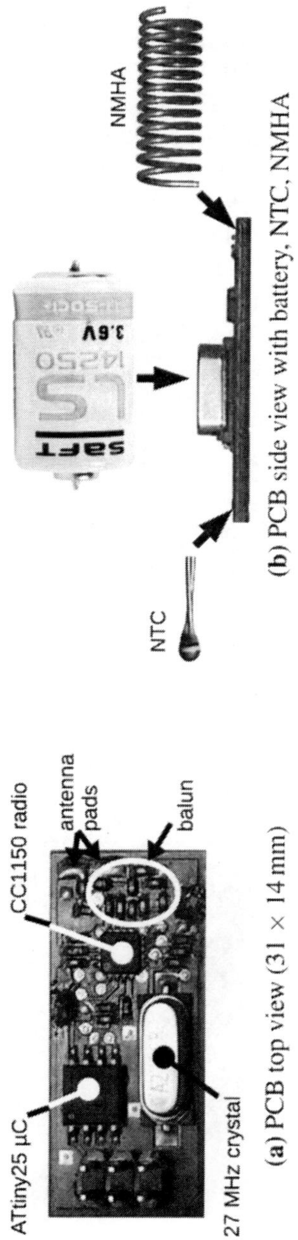

Figure 3. Sensor node prototype: PCB for laboratory tests (**a**); and prepared for field tests with manually soldered NTC , normal mode helix antenna (NMHA) and battery (**b**); the RF performance is limited by the balun layout (a) and the proximity to the antenna of ferromagnetic crystal case and battery seal (b).

The average standby current of the sensor node is about 4.7 μA, largely due to the watchdog timer. It samples the temperature once per second with an NTC transducer (discussed in Section 5.1), keeping both the microcontroller and the transducer active for less than 0.05% of the time. The former consumes about 600 μA in active state and the latter a few tens of microamperes; thus, their contribution to node average energy consumption is negligible. The transmission of one packet consumes about 27 mA for less than 0.3 s. Thus, in normal operation (sending one heartbeat packet per hour, as discussed in Section 4.1), the average current consumption is less than 7 μA, and the node theoretical lifespan with a 1 Ah 1/2-AA size lithium battery exceeds 16 years.

4.3. Gateway Node Design

The gateway prototype shown in Figure 4 is an evolution of an earlier prototype using a small ATmega324P microcontroller (32 KiB FLASH, 2 KiB SRAM, 1 KiB EEPROM) that was continuously relaying the data from the sensor nodes to the application server using the on-board general packet radio service (GPRS) modem, with almost no power management. It helped with testing the sensor node and server prototypes and with developing the modem driver and the recovery mechanisms from various GPRS connection errors.

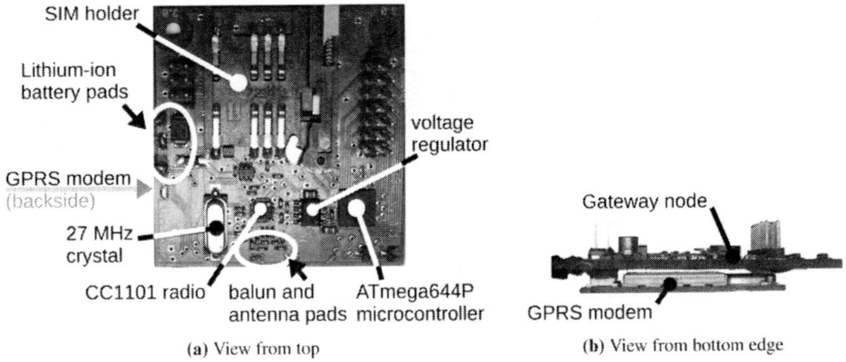

Figure 4. Top view (**a**) and bottom-edge view (**b**) of the gateway node prototype (50 × 50 mm). The layout of the balun (a) limits the RF performance and the battery charger of the modem is not efficient for energy harvesting.

The evolution shown in Figure 4 uses an ATmega644P microcontroller (64 KiB FLASH, 4 KiB SRAM, 2 KiB EEPROM), which has more SRAM and allows more extensive sensor node message aggregation. It uses a TI CC1101 transceiver for in-field communications, compatible with the sensor node transmitter. The GPRS modem (on the back side of the board; seeFigure 4b) is a Siemens TC65i with an embedded TCP/IP stack and a charge regulator for lithium-ion batteries, which can be used for energy harvesting.

It is worth noting that the gateway node prototype is cost- and energy-effective, since it uses hardware resources that are typical for a low-end WSN sensor node (except for the GPRS modem).

5. PLATFORM PREPARATION FOR THE FIELD TEST

While the platform prototype was still in the early stages of development, it became unexpectedly necessary to prepare all of its components (the sensor and gateway nodes and the server) for an extended two and a half month field test (end of June to early September). The test was set to start in less than two months using 50 sensor nodes and two gateways deployed in an olive tree field in the south-west of Greece. In the following, we will present how we selected, developed and assembled in this short time a platform prototype with the most important features to test in the field. We will also show how we used the results and experience from the test to refine and enhance platform functionality and reliability.

The time to field test was too short to design, produce and test new sensor nodes, and we decided to adapt the functionality and ruggedize the node prototypes designed for laboratory tests (described in Section 4). As expected, this adaptation has led to several issues during the test, but also fostered a thorough analysis of the platform operation before, during and after the test, ultimately helping us to significantly improve its reliability and cost, as we will show below.

Because the test field was located far away (nearly 2000 km), on-site maintenance would have been costly and time-consuming. On the other hand, since the components of the platform were not developed with the quality necessary to withstand an extended environmental exposure in field, we expected to encounter hardware issues during the test.

Thus, we made a development priority to add the necessary monitoring features to collect enough data about defects and anomalies that may manifest during the test, in order to understand their causes and effectively help with improving the future versions of the platform. Most of these features would be needed anyway, since QoS control is always important for long-term outdoor environmental monitoring platforms, especially when monitoring for potentially dangerous events, like wildfires, as was discussed in Sections 3.1.3 and 3.2.3.

Based on the analysis in Sections 3.1.2 and 4.1, we chose a star network topology and a very simple communication protocol, because they can considerably simplify the sensor nodes: reduce the requirements for hardware resources, the size and complexity of the communication protocol stack, the energy consumption and the cost. All of these are important specifications for pervasive monitoring platforms and, most importantly, they simplify network deployment to a level where it can be efficiently done by application domain experts or low-skill workers.

With the above considerations, we split the network into independent cells, as shown in Figure 5.

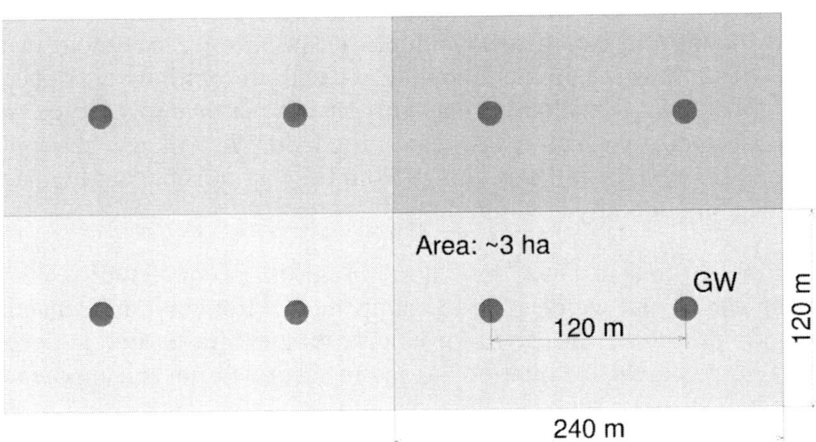

Figure 5. The network is made of independent cells with a star topology inside (the distances assume a 200m sensor communication range). The sensors in each cell are covered by two gateways, which can independently contact the server to ensure redundant field message aggregation and upload. Larger areas can be covered by joining elementary cells.

The sensors in each cell are distributed with the density defined by the domain experts based on terrain and vegetation characteristics (e.g., every 20 m) and are covered by two gateways. Each gateway can independently contact the server in order to ensure redundant field message aggregation and upload. Larger application areas can be covered by joining elementary network cells, as shown in Figure 5.

Although with a link budget of, e.g., 110 dBm, the communication range in the 433-MHz band can well exceed 10 km in optimal propagation conditions, the actual range in forest (or in the test field) conditions may be much shorter. Thus, the distances in Figure 5 assume a sensor communication range of 200 m, and the test field shown in Figure 6 of about 195 × 45m can be adequately covered by one cell.

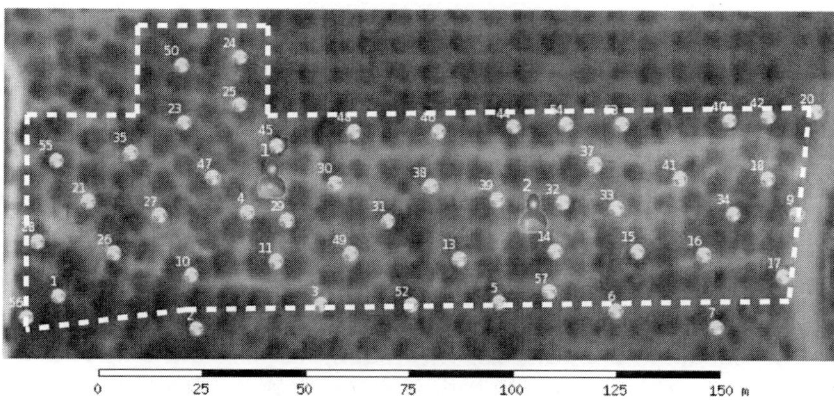

Figure 6. The test field is about is about 195 × 45m (main body). Trees appear as darker spots on the light-colored soil. There are two gateways and 50 sensor nodes (round dots; some are placed outside the field due to GPS positioning errors).

This layout simplifies network deployment (presented in Section 6.1) and ensures full redundancy for field data collection, storage and communication to the server.

5.1. Preparation of the Sensor Node for the Field Test

The sensor node prototype described in Section 4.2 was designed for laboratory tests. It lacked several features for field tests, such as a temperature transducer, its interface circuit and a microcontroller input to read the transducer. Thus, the node schematic (shown in Figure 7a) was manually changed directly on the PCB, as shown in Figure 7b. R3 (an NTC) and R2 were connected as a voltage divider that was supplied only during the readings (to reduce node energy consumption) by reusing the SPI output pin PB1 as a software-controlled voltage source. At the same time, PB4 was reconfigured as an ADC input to read the transducer, cutting its connection to the radio module (which was needed only for tests).

These hardware and software changes allowed the sensor node to perform its main functions, monitoring the ambient temperature and notifying about abnormal high levels or too steep increases. The sensor nodes were programmed to measure the temperature once per second and to send specific alert messages when levels highly correlated with wildfires are detected.

The sensor node has to support also a fast and error-proof field deployment procedure, as discussed in Section 3.1.3. Even if the field test was using just a few dozen sensors, it was worth defining and testing from the beginning an effective field deployment procedure suitable for large applications, with hundreds or thousands of nodes that are deployed by personnel with no special WSN training.

These three items are especially important during the deployment of homogeneous sensors:

1) the ID of the sensor node;

2) the quality of the reception of the sensor node messages by the gateways;

3) the field coordinates of the sensor node deployment spot.

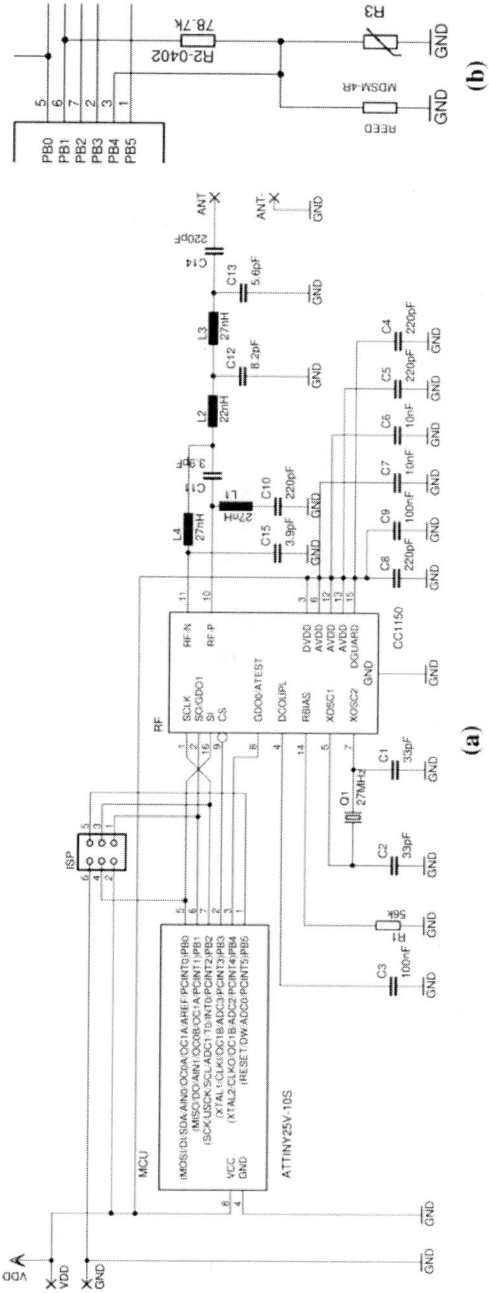

Figure 7. Schematic of the sensor node prototype for laboratory testing (**a**); issues fixed (**b**): (1) add a temperature transducer with an interface circuit and (2) add a means to switch the sensor to deployment mode.

The node deployment procedure should minimize human input, because the operators may only have limited training and time. Thus, the automation of the procedure can significantly reduce deployment errors and cost.

For an efficient field deployment, it is very important to be able to selectively address the node to be installed while other nodes are within comparable short distance (e.g., in the operator's backpack). Relying on the operator to input the node ID can be very time consuming and error prone, so an inexpensive (component cost and energy) near field communication with the node to install is necessary.

For this purpose, we added a reedswitch in parallel with the NTC (see Figure 7b), which can be activated using a magnet from a distance up to about 2 cm. When activated, the ADC will read 0V as the transducer output, which can be easily recognized, and temporarily switch the node to installation mode. In this mode, the node sends a few messages prompting the gateways in range to generate special acknowledgments that include the quality of the signal received from the node and the node ID. These data can be displayed on a handheld device to provide immediate feedback to the operator on the quality of the node installation position under evaluation. The handheld device can also determine the position of the node using an on-board GPS receiver and send all of the data to the server once validated by the operator.

This node installation flow requires minimum input from the operator, specifically:

- trigger the deployment mode of the node by rubbing a magnet on its case. This is an intuitive and error-proof way to address the node under installation;
- read on a handheld device the quality of the current deployment position;
- tap a button on the handheld display to validate the deployment and send the data to the server.

Because of the short time available to prepare the field test, we implemented this flow only partially. We mounted the reed switches on the sensor nodes and changed the gateway protocol to provide the reception quality in the acknowledgments. The handheld device was implemented using a modified gateway that was displaying only the lowest reception level of the gateways in range on three LEDs.

During normal operation, the sensor node periodically tests the on-board temperature transducer given its importance for the detection of the fire. The node can detect short-circuits of the transducer (decoded as installation requests, as described above) and open-circuits. The node notifies to the gateways both faults in order to propagate them to the server which warns the operators about the fault.

Besides recognizing the extreme cases of short- and open-circuit, the selection of the temperature thresholds for alerts is very important. Low thresholds ensure timely alerts, but they also increase the sensor susceptibility to false alerts triggered by natural phenomena, such as sensor heating by direct exposure to sunlight. These considerations are especially important for fields with high solar exposure, as is the test field during summer time. After some testing, we decided that the fire alert threshold can be set to 60 °C and the threshold for warning of possible fire to a temperature increase rate of 1 °C/s.

A normal mode helix antenna (NMHA) was soldered to the antenna pad on the right of the PCB, as shown in Figure 3b. The NMHA is made of about 12 turns on an internal diameter of ϕ 10mm with a ϕ 1mm copper wire and has about a 21.5mm length. The number of turns and the length were experimentally tuned to maximize the RF radiation in deployment-like conditions.

The sensor node package designed for the field test is shown in Figure 8. It holds the node PCB shown in Figure 8b with a half-AA battery soldered on top, an NTC transducer to the left and the NMHA antenna mounted axially to the right. The programming head on the left of the PCB was removed, and everything was sprayed with a protective resin against moisture and insects. The aperture on the left side of the package improves the thermal contact between the NTC and the ambient air. The node is mounted vertically on trees, at a height of 0.5 m, with the aperture pointing downwards for a good contact with the hot air rising from ground fires, which are notoriously difficult to detect using other techniques.

5.2. Preparation of the Gateway for the Field Test

To reduce the cost, the gateway node prototype for laboratory tests (described in Section 4.3) is implemented using resources that are typical for a WSN sensor node. The prototype can be used for the field test, although it is not optimized for low energy, as we will show below.

(a) (b)

Figure 8. Sensor node package hosting the PCB, a half-AA battery and the normal mode helix antenna (NMHA) antenna. The aperture on the left improves the NTC thermal contact (**a**); fixed on trees with screws (**b**).

The resources of the microcontroller are sufficient for the gateway tasks. However, more SRAM would allow it to aggregate additional field messages, reducing the frequency of connections to the server and the energy consumption.

The transceiver for in-field communication is used to receive messages from sensor nodes and to communicate with peer gateways for health checks and to increase the availability of service, e.g.:

- in case of failure of the GPRS link of a gateway:
    - it can send its field messages to peers, so that they forward them to the server;
    - send self-diagnostic data to the server through peers to provide the remote operators the necessary information for corrective actions;
- failures of the field receiver would prevent the gateway from receiving hello messages from peers; thus, the gateway would eventually declare all peers "missing"; however, since the peers would regularly contact the server without reporting errors, this condition can be detected and appropriately flagged on the server;

- failures of the field transmitter would prevent the peers from receiving the gateway hellos, which they will report to the server; the faulty gateway itself would not report any anomaly.

Gateways require much more energy than sensor nodes due to the continuous LPL of the sensor node channel (to support the random MAC) and energy consumption of the GPRS modem. Although the latter is turned off most of the time, the average energy consumption of the gateway is still too high to allow its long-term operation from a reasonably-sized primary cell. Thus, we installed solar panels on the gateway for energy harvesting, since the test field receives abundant sunlight.

Considering the above, the gateway was embedded in a birdhouse, as shown in Figure 9a.

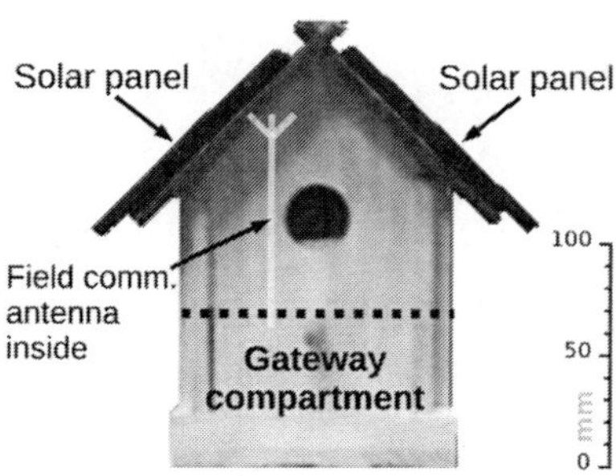

(a) The birdhouse has solar panels on the roof and the gateway, battery and GPRSmodem in a sealed package at the bottom. The antenna between the solar panels has low RF performance.

(b) Gateway 2 is deployed in the field high above the ground to improve its RF coverage. The ground of the test field is clean, facilitating navigation and sensor deployment and retrieval.

Figure 9. Gateway package elements (**a**) and field placement of Gateway 2 (**b**).

The solar panels for energy harvesting were placed on the roof, and the internal space of the birdhouse was divided horizontally into two compartments using a wooden separator. In the lower one were placed the gateway, the rechargeable battery and the GPRS modem in a sealed plastic package. The antenna of the GPRS modem was left inside the plastic package, because the test field had good cellular coverage. The antenna for in-field UHF communications, a $\lambda/4$ whip, was kept inside the birdhouse for aesthetic reasons. As expected, this position reduced the RF sensitivity, as we will discuss below.

5.3. Preparation of the Server for the Field Test

The structure of the application server is shown in Figure 10 and has two major parts. One implements the core functions, such as listening on a socket for incoming connections from the gateways, decoding, saving and replicating field messages and alert management. The other implements the self- and external service checks, such as network connectivity, peer server, local database and SMS providers for alerts.

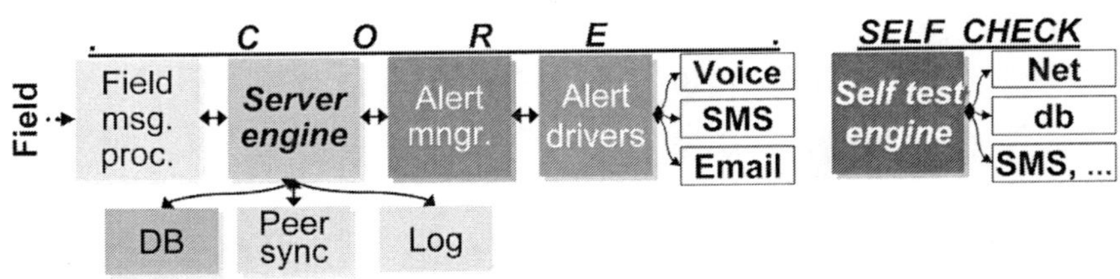

Figure 10. The application server structure is split into core services and a self-test engine.

For the field test, the application server was updated to support the extensions of the gateway protocol, such as to send the receive acknowledgments to gateways to prevent field data loss, support for new message types for field device health semantics, session initiation and termination and time synchronization with the gateways. The user interface was extended to effectively display many sensor nodes with multiple health states, to improve responsiveness and usability and provide several operator authorization levels.

6. FIELD TEST OF THE WSN PLATFORM

Field deployment was scheduled on two days. On the first day, the field planning that was prepared in advance (remotely) was checked, followed by the actual field deployment. During the second day, the operation of the system was checked in the field for final tunings before commissioning to field test. Larger deployments may take more effort and time to complete, but they can follow the same sequence of operations.

6.1. Field Deployment of the WSN Nodes

Because the system was experimental and the field was rather small, we decided to monitor all communications between the nodes during field deployment from a fixed location near the field, using a modified gateway that was programmed to decode and display the messages as they were sent in the field. To detect abnormal behaviors during deployment and to improve the procedure, the meaning of these messages was manually compared with the state of the field displayed by the server, which was built using the field information received from gateways.

6.1.1. Field Deployment of Gateway Nodes

The gateways were deployed in the locations shown in Figure 6. Gateways are always deployed before sensor nodes for star topologies to allow checking the connectivity of the nodes during installation.

While the sensor node deployment procedure described in Section 5.1 provides the means to check the quality of its connectivity within the network, no such provision was made for the gateways, because of the short time we had to prepare the field test. The assessment of the quality of a gateway deployment would include cellular coverage for the GPRS modem, light exposure for energy harvesting and the coverage of the sensor nodes in the field.

For this test deployment, we checked manually in the server log that the gateways were able to successfully connect. Furthermore, we monitored the field messages in real time for reports of success (or errors and retries) of the connection steps performed by the gateways: register with the cellular operator, access the GPRS service and open a socket with the server over the Internet.

The deployment places for the gateways were chosen so that one solar panel was roughly directed towards south or on the east-west axis, and the sunlight was reaching them without major obstructions. Figure 9b shows an overview of part of the test field and the deployment position of Gateway 2.

6.1.2. Field Deployment of Sensor Nodes

Once the gateways were in place and operational, we started to deploy the 50 sensor nodes in the locations shown inFigure 6 with the procedure described in Section 5.1. Due to the limited time we had to prepare the field test, the deployment procedure was only partly automated. While the deployment mode of the sensor nodes can be quickly activated using a magnet, the operator would still need to manually record the node ID and position coordinates read from a handheld GPS. These operations proved to be time-consuming and error-prone and required us to make several passes through the field to recheck node IDs and positions. Moreover, as can be seen in Figure 6, some positions were still slightly out of the field (e.g., Sensor Nodes 2, 6, 7) due to the limited accuracy of the GPS receiver.

Using the node deployment procedure presented in Section 5.1 (only partly automated, as we have mentioned), we needed about 4–5 min to install one sensor node in a relatively simple field (see Figures 6 and 9b): the trees were well spaced and organized in rows; most ground vegetation and obstacles were cleared; and the GPS signal was received without the typical attenuation from dense tree crowns.

Nevertheless, it took us about four hours to deploy all 50 sensor nodes. This means that node deployment is generally effort intensive and can significantly increase the cost of larger applications.

6.2. Issues during Field Deployment

6.2.1. RF Range Issues

Although the size of the test field was relatively small compared to the communication range in ideal conditions of about 17 km for the 433-MHz band with a 110 dBm link budget, we noticed during deployment that some nodes at more than 50–70m from gateways were received little or not at all.

Several design decisions reduced the communication range. First, the placement of the gateway antenna for field communications between the solar panels of the birdhouse (see Figure 9a) reduced its efficiency. We expected this, but aesthetic considerations prevailed, since we had not the time to measure the actual loss before the test. After the test, we measured about a 24–26 dBm loss, a penalty that led us to install external sockets for all aerials in the new birdhouse project.

After the test, we discovered two additional RF attenuation causes on the PCB of the sensor node. As shown in Figure 3b, the antenna is just a few millimeters away from two large ferromagnetic items: the battery side seal and the case of the quartz crystal resonator. The typical loss of the NMHA is about 6–8 dBm, but these two elements combined were adding about a 10 dBm loss. Thus, in the next version of the sensor node PCB, we will use a different case for the crystal resonator and move the battery away from the antenna to reduce its RF attenuation.

The layout of the baluns of both the sensor node and the gateway (shown in Figures 3a and 4a) was compacted to reduce the size of the nodes, size being especially important for the sensor node. One of the purposes of the node prototype was to determine the loss of the compact balun layout, which we determined after the test to be around 3 dBm. Thus, we accepted the larger size of a straight balun layout.

The package of the sensor node (shown in Figure 8) was introducing an attenuation of about 2 dBm, but this cannot be easily avoided.

Another attenuation cause was due to terrain geometry. The test field shown in Figure 6 had a very gently slope at the bottom-left corner, almost imperceptible to the eye. To increase the sensitivity to heat released by ground fire, we installed the sensor nodes low on the tree trunks, about 0.5m above the ground. The gateways, on the other hand, were placed higher in the trees to improve both energy harvesting from sunlight and RF reception from low-placed sensor nodes. However, Node 1 and, even more, Node 56 were attenuated too much by this small terrain undulation, which was easily discovered by the checks performed during deployment. Thus, we decided to install the affected nodes higher, at 1m from the ground. Nevertheless, this incident shows that even small terrain undulations may restrict the coverage of the gateways, especially for distant nodes, and this should be taken into account when choosing the positions of the gateways in the field. We also decided to develop a repeater node that can be deployed to improve the gateway coverage of the field.

In general, the deployment positions of the gateways on hilly terrain should be carefully analyzed during planning. Nevertheless, there can always be propagation issues that are discovered only during deployment (e.g., due to fading from multipath propagation or obstacles). Some of these issues can be fixed by changing the position of the sensors (a different position on the tree trunk, higher from the ground or on a nearby tree). In other cases, we can always add a gateway to cover these areas or, even better cost-wise, a repeater device.

6.2.2. Temperature Issues for Sensor Nodes

As the deployment of the sensor nodes progressed and the Sun approached the zenith, some of the sensor nodes waiting to be deployed started to send fire alerts. Apparently, this was because the nodes were transported in a dark color backpack which excessively heated them while lying idle in direct sunlight (the test field had only a few patches of shadow, as can be seen in Figure 9b).

Rising the alert threshold of the sensor nodes would avoid false alerts, but it would also reduce the detection sensitivity of fires, delaying the alerts. Since timely alerting is an important feature of the system, we chose to revise the transport and deployment procedure to prevent sensor node overheating (e.g., by using lighter colors for packs during transport and avoiding extended exposure to sun heat).

6.3. Long-Run Test of the WSN Platform

After all nodes were deployed and the operation of all platform components was checked, the field test started for two and a half months.

6.3.1. Gateway Issues during the Long-Run Test

Post-deployment checks missed a very important fault. The battery charging circuit of the GPRS modem of Gateway 2 failed during deployment and was not charging the battery. We discovered this too late, after we left the field, because we were monitoring the battery voltage (measured by the GPRS modem), and the potential *vs.* charge relationship, shown inFigure 11, has an extended, almost flat plateau that makes it difficult to determine changes in battery charge from voltage readings.

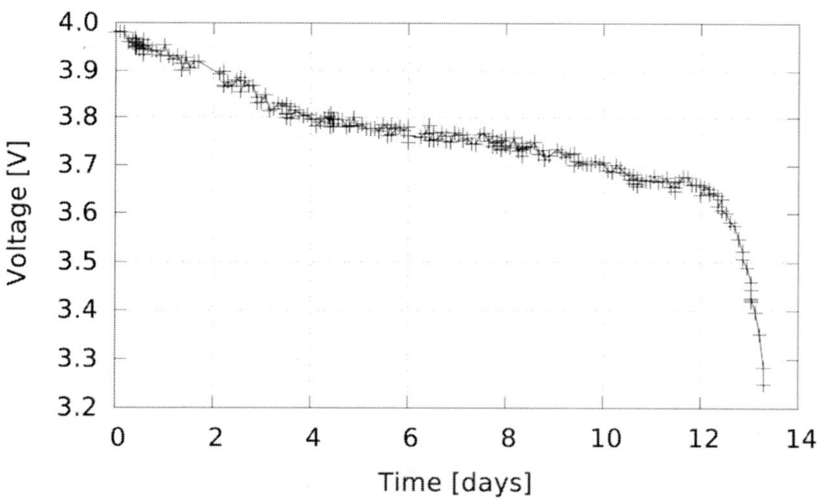

Figure 11. Lithium-ion battery voltage discharged by a gateway that contacts the server every 15 min to transfer 50 messages. It has an almost flat plateau, just below 3.8V.

The battery charger of the modem of Gateway 2 may have failed because of overheating during deployment due to a combination of the following three factors. First, Gateway 2 was the first deployed, thus it received more sun heat, which increased its internal temperature. Second, it received and forwarded to the server most of the false fire alerts generated by the sensor nodes, which additionally increased the temperature of the modem. Third, since the solar panels were well illuminated, they were charging the battery, thus increasing the temperature of the charging circuit even more.

As Gateway 2's battery progressively discharged, we monitored the effectiveness of the built-in in-field redundancy. The GPRS modem does not work below 3.2V, but the rest of the gateway (microcontroller and field radio) operates until the battery is completely discharged (lithium-ion batteries stop working around 2.7 V). When battery voltage dropped below the low threshold of the GPRS modem and it ceased to work, Gateway 2 continued to operate as a repeater, forwarding to its peer the messages received from the sensor nodes in its range. Gateway 1 aggregated them with its own messages and sent them all to the server, maintaining the redundant coverage of field nodes, as expected. When Gateway 2 eventually stopped working completely, Gateway 1 detected the new fault condition and notified the operators through an error message sent to the server.

Another temperature-related issue was due to the limited range allowed for charging the lithium-ion batteries, 0–45 °C. Open nature temperatures can exceed this range, thus blocking battery recharging and, eventually, reducing its lifetime. These limits make lithium-ion rechargeable batteries less suited for open space environmental monitoring.

During our field test, high daytime ambient temperature combined with heat from gateway direct sun exposure and heat released during GPRS communication and battery charging were often blocking battery charge during the central hours of hot days. Figure 12a shows how the internal gateway temperature met or exceeded 45 °C on most days over a two-week period. However, this did not significantly affect the operation of the gateway, since hot days are typically also very sunny, thus providing sufficient energy for recharging. Nevertheless, heat dissipation was not adequate for this gateway prototype. The gateway package should be adapted to the climate of the monitored field.

Figure 12b shows how low temperature may break the synchronous serial communication between the microcontroller and the GPRS modem (approximately between Days 6–13 and 43–45 in the figure). However, since these failures correlate loosely with temperature, we could not rule out other causes, such as condensing humidity inside the gateway or inside the GPRS modem, until we reproduced the ambient conditions using a climatic chamber after the field test. The log collected locally on the affected microcontrollers indicated serial communication errors that were resolved by recalibrating the internal oscillator of the microcontroller from an accurate reference (e.g., a scaled-down frequency of the field radio crystal resonator).

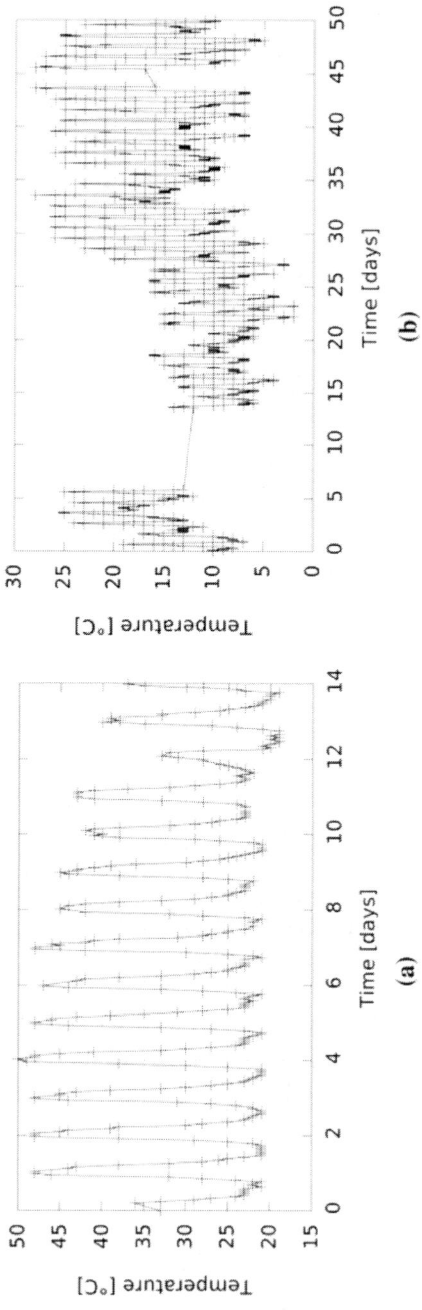

Figure 12. Gateway temperature can exceed 45 °C, the maximum allowed for lithium-ion battery charging (**a**), while the internaloscillator drift can break the synchronous serial communication with the GPRS modem at low temperatures (Days 6–13 and 43–45) (**b**).

6.3.2. Sensor Node Issues during the Long-Run Test

Thermal expansion mismatch between the PCB and the NTC leads generated repeated mechanical strains due to cyclic temperature variations of 15–20 °C between day and night. These eventually led to fatigue cracking of NTC solder joints on some sensor nodes, which, in turn, led to intermittent contacts, especially during morning and evening hours, when the temperature changed most rapidly. The sensor node decoded the imperfect contacts as fast ambient temperature changes, and notified these right away to the gateways as wildfire warnings (see Section 5.1). The gateways handled these warnings the same way as fire alerts, forwarding them immediately to the server to promptly notify the operators.

Since the GPRS modem is one of the highest energy drains of the gateway, keeping it powered up for a few hours in the morning and evening every day to forward these warnings eventually consumed more energy than the harvesting capacity of the solar panels and discharged the battery of the only gateway left. Its failure required a maintenance operation to restore both gateways to full functional state and to replace the faulty sensor nodes.

Such sensor node faults can happen also in large deployments that should run unattended for many years. Thus, the system should be able to exclude the network effects of faulty nodes before they can create larger problems that may lead to (partial) loss of service. For this purpose, we decided to provide the gateways with a list of sensor node IDs that should be ignored and which can be set remotely by the operators when needed. Such a list would have helped us avoid a costly field trip for maintenance during the test.

Last, but not least, all systems powered by harvested energy (which may fluctuate widely reaching very low levels) should have brown-out detection (BOD) enabled to avoid entering non-functional states when the battery voltage is too low. Because the BOD was not enabled on the gateway prototypes, Gateway 1 failed several times to return to normal operation after periods of insufficient energy.

7. CONCLUSIONS AND FUTURE WORK

In situ pervasive environmental monitoring for wildfires has important advantages, but the relatively high area cost may limit its use. We have shown in this paper several domain-specific platform optimizations that reduce the cost and increase its economic viability.

Sensor node hardware and a communication protocol optimized for application requirements can exceed 10 years of life time using a small, low-cost primary cell. Long node life significantly reduces the cost of the scheduled maintenance of the network, which can be important for large networks. The nodes and the communication protocol also speed up field deployment. The deployment procedure prevents most human errors, significantly reducing the cost, especially for networks with a large number of nodes.

Field experiment preparation and unfolding highlighted several aspects that can improve the cost and reliability of the network components, their deployment, operation and maintenance.

Long-term direct node exposure to climatic conditions increases the failure probability. Additionally, large networks are more likely to experience node failures. As we have seen during the field experiment, failed nodes may adversely affect network operation or strain the resources of the gateways in range, increasing maintenance cost. Thus, it is important to increase node and network robustness during design, production testing, functional and self-test specifications and run-time monitoring.

As we have shown, node hardware design and production testing were not adequate for the field experiment and led to several node failures. Accumulation of mechanical stress should be reduced by design and during production, since it compounds, e.g., with the stress from daily and seasonal temperature variations or unforeseen mechanical interaction with wild animals. Manual soldering should be avoided, since it can inadvertently create strains in solder joints in the interaction with component leads.

Faulty nodes should be detected as soon as possible to reduce production costs, e.g., by running specific self-test programs at the end of the production line.

Lithium primary batteries have shelf lives in excess of 10 years. Compared to ambient energy harvesting techniques suitable for open nature, primary cells can reduce node cost, increase reliability and reduce node deployment complexity. These cells can supply very low power nodes cost effectively and maintenance free for more than 10 years, a life time suitable for many WSN applications.

Brown-out detectors may consume too much energy and should not be needed for long lifetime nodes with soldered batteries. However, it is important to reset the nodes right after the battery is soldered to avoid leaving them in abnormal states that may excessively drain the battery.

Sensor node functions and a communication protocol carefully tailored to the application requirements can bring important cost reductions. For instance, transmit-only nodes in star topologies can reduce the protocol stack and relax the accuracy requirements for on-node timing. In turn, these reduce microcontroller resources and energy consumption, significantly lowering the cost of components, battery and node maintenance. However, without a receiver, the nodes cannot create mesh topologies to improve network robustness and to reduce transmit power. Thus, the cost and benefits of such optimizations should be evaluated in the context of the specific application or application class.

Periodic node self-testing and automatic error recovery increase the reliability and the QoS of the monitoring platform, reducing maintenance requirements and cost. In case of critical failures (e.g., failing contact with the NTC), the nodes may disable themselves after reporting the fault to avoid reporting false events or impairing the operation of the network or the gateway nodes, as we have witnessed during the field experiment.

Small sensor node physical dimensions and protection from environmental conditions may reduce the RF efficiency in several ways. For instance, the antenna may end up too close to large ferromagnetic bodies, like the battery side seal or the case of the quartz crystal resonator in our experiment. While we could use a different case for the crystal resonator to avoid changing the node geometry, we had to increase node length to leave enough distance between the antenna and the battery seal. Furthermore, protecting the antenna inside the node package reduced its efficiency by 2 dB, but we found no cost-effective way to avoid it. However, component cost is less important for gateways, and we plan to install their antennas clear of the package or solar panel shielding, which have markedly reduced their performance during the field test.

Changes in gateway behavior in the case of failures (e.g., GPRS link or field radio receiver failures) need to be detected at the gateway or network level and reported to network operators to schedule corrective actions. Remote configuration or reprogramming should be available, at least for the gateways, in order to be able to remotely add functions, correct errors or protect the network against occasional faulty node behavior. As we have seen during the field experiment, this feature would have spared us an unscheduled maintenance, whose cost can be important for remote fields or large networks.

Nodes that store field messages before forwarding (e.g., gateway nodes) should implement suitable in-field redundancy to avoid loss in case of self-reset or failure. Additionally, we plan to implement a save-restore function on gateways to preserve node data across watchdog-triggered resets.

Gateway nodes handle higher energies, which can generate significant heat, both during harvesting and consumption peaks (e.g., during GPRS data connections with the server). Thus, as we have seen during the field experiment, an adequate thermal dimensioning considering the climate and the internal and external heat sources is important. For some applications, we may consider other energy accumulation techniques instead of the widely-used lithium-ion batteries, which have limited charge-discharge cycles and temperature restrictions for charging and operation.

We also consider redesigning the power supply of the gateways to automatically switch to a primary cell whenever the harvested energy reserves are too low, to increase the node QoS. Unlike the sensor nodes, which are reliably supplied by the battery soldered on-board, the energy reserves of the gateways can fluctuate due to unpredictable harvesting and consumption patterns. Moreover, the rechargeable batteries can be changed when they loose storage capacity. Hence, the use of the embedded brown-out detector can considerably improve gateway stability.

The field experiment showed us that the deployment of the network can be very time-consuming, even for networks with just a few dozens of nodes. Thus, the procedure should be simple and automated as much as possible to be accessible to domain operators, to reduce human errors and cost. Nevertheless, it should allow one to thoroughly check the quality of node installations, since RF propagation in open nature can be unpredictable, not the least due to small and difficult to notice terrain features. We plan to improve the user interface of the handheld installation device, automate position collection and data reporting and extend it to the installation of the gateways. For the latter, it should assess cellular coverage for the GPRS modem, light exposure for energy harvesting and the coverage of the sensor nodes in the field. We also decided to develop a repeater node that can improve gateway coverage of the field in difficult propagation conditions.

WSN platform design, development and maintenance has a high cost, which can be recovered faster if the platform can be reused for several related application domains. While the sensor nodes do not have sufficient resources to support an IP-based protocol, the gateways are IP-enabled, albeit they are sporadically connected. The server, however, is always accessible over the Internet, and in future versions, it can implement suitable interfaces (e.g., based on OGC® Sensor Web Enablement [91–94]) to facilitate the discovery and access to sensor and field data by authorized IoT applications.

ACKNOWLEDGEMENTS

Most part of this work was done in collaboration with Minteos (Minteos s.r.l. http://www.minteos.com/) and the Artemis WSN-DPCM project. The company contributed mostly to application specification, field testing and production.

CONFLICTS OF INTEREST

The authors declare no conflict of interest.

REFERENCES

1. Ashton, K. *That "Internet of Things" Thing. Expert View*; RFID Journal, 2009. Available online: http://www.rfidjournal.com/article/view/4986 (accessed on 17 April 2015).

2. Oppermann, F.; Boano, C.; Römer, K. A Decade of Wireless Sensing Applications: Survey and Taxonomy. In *The Art of Wireless Sensor*

Networks; Ammari, H.M., Ed.; Signals and Communication Technology, Springer: Berlin/Heidelberg, Germany, 2014; pp. 11–50.

3. Romer, K.; Mattern, F. The design space of wireless sensor networks. *IEEE Wirel. Commun.* **2004**, *11*, 54–61.

4. Corke, P.; Wark, T.; Jurdak, R.; Hu, W.; Valencia, P.; Moore, D. Environmental Wireless Sensor Networks. *IEEE Proc.***2010**, *98*, 1903–1917.

5. Rault, T.; Bouabdallah, A.; Challal, Y. Energy efficiency in wireless sensor networks: A top-down survey. *Comput. Netw.* **2014**, *67*, 104–122.

6. Strazdins, G.; Elsts, A.; Nesenbergs, K.; Selavo, L. Wireless Sensor Network Operating System Design Rules Based on Real-World Deployment Survey. *J. Sens. Actuator Netw.* **2013**, *2*, 509–556.

7. Harrop, P.; Das, R. *Wireless Sensor Networks 2010–2020*; Technical Report; IDTechEx Ltd: Cambridge, UK, 2010.

8. Rousselot, J.; Dallemagne, P.; Decotignie, J. Deployments of wireless sensor networks performed by CSEM. Proceedings of the COGIS 2009—COGnitive Systems with Interactive Sensors Conference, Paris, France, 16–18 November 2009.

9. Kumagai, Y.; Carroll, M.S.; Cohn, P. Coping with Interface Wildfire as a Human Event: Lessons from the Disaster/Hazards Literature. *J. For.* **2004**, *102*, 28–32.

10. Stephens, S.L. Forest fire causes and extent on United States Forest Service lands. *Int. J. Wildland Fire* **2005**, *14*, 213–222.

11. Genton, M.G.; Butry, D.T.; Gumpertz, M.L.; Prestemon, J.P. Spatio-temporal analysis of wildfire ignitions in the St Johns River water management district, Florida. *Int. J. Wildland Fire* **2006**, *15*, 87–97.

12. Arvai, J.; Gregory, R.; Ohlson, D.; Blackwell, B.; Gray, R. Letdowns, wake-up calls, and constructed preferences: People's responses to fuel and wildfire risks. *J. For.* **2006**, *104*, 173–181.

13. Leone, V.; Lovreglio, R.; Martín, M.P.; Martínez, J.; Vilar, L. Human Factors of Fire Occurrence in the Mediterranean. In *Earth Observation of Wildland Fires in Mediterranean Ecosystems*;

Chuvieco, E., Ed.; Springer: Berlin Heidelberg, Germany, 2009; pp. 149–170.

14. Shroder, J.F.; Paton, D. *Wildfire Hazards, Risks, and Disasters*; Elsevier: Philadelphia, PA, USA, 2014.

15. Salis, M.; Arca, B.; Ager, A.; Fois, C.; Bacciu, V.; Duce, P.; Spano, D. Extreme Wildfire Spread and Behaviour: Case Studies from North Sardinia Italy. Proceedings of the EGU General Assembly Conference, Vienna, Austria, 22–27 April 2012; Volume 14.

16. Hefeeda, M.; Bagheri, M. Forest fire modeling and early detection using wireless sensor networks. *Ad Hoc Sens. Wirel. Netw.* **2009**, *7*, 169–224.

17. Stipaničev, D.; Štula, M.; Krstinić, D.; Šerić, L.; Jakovčević, T.; Bugarić, M. Advanced automatic wildfire surveillance and monitoring network. Proceedings of the VI International Conference on Forest Fire Research, Coimbra, Portugal, 15–18 November 2010; pp. 15–18.

18. Šerić, L.; Stipaničev, D.; Štula, M. Observer network and forest fire detection. *Inf. Fusion* **2011**, *12*, 160–175.

19. Ma, L.; Wu, K.; Zhu, L. Fire Smoke Detection in Video Images Using Kalman Filter and Gaussian Mixture Color Model. Proceedings of the International Conference on Artificial Intelligence and Computational Intelligence, Sanya, China, 23–24 October 2010; Volume 1, pp. 484–487.

20. Tsiourlis, G.; Andreadakis, S.; Konstantinidis, P. SITHON: A Wireless Network of in Situ Optical Cameras Applied to the Early Detection-Notification-Monitoring of Forest Fires. *Sensors* **2009**, *9*, 4465–4482.

21. Arrue, B.; Ollero, A.; Matinez de Dios, J. An intelligent system for false alarm reduction in infrared forest-fire detection. *IEEE Intell. Syst. Their Appl.* **2000**, *15*, 64–73.

22. Yoon, S.H. An intelligent automatic early detection system of forest fire smoke signatures using Gaussian mixture model. *J. Inf. Process. Syst.* **2013**, *9*, 621–632.

23. He, L.; Li, Z. Enhancement of a fire detection algorithm by eliminating solar reflection in the mid-IR band: Application to AVHRR data. *Int. J. Remote Sens.* **2012**, *33*, 7047–7059.

24. Çetin, A.E.; Dimitropoulos, K.; Gouverneur, B.; Grammalidis, N.; Günay, O.; Habiboğlu, Y.H.; Töreyin, B.U.; Verstockt, S. Video fire detection—A review. *Digit. Signal Process.* **2013**, *23*, 1827–1843.

25. Vodacek, A.; Kremens, R.L.; Fordham, A.J.; Vangorden, S.C.; Luisi, D.; Schott, J.R.; Latham, D.J. Remote optical detection of biomass burning using a potassium emission signature. *Int. J. Remote Sens.* **2002**, *23*, 2721–2726.

26. Dennison, P.E. Fire detection in imaging spectrometer data using atmospheric carbon dioxide absorption. *Int. J. Remote Sens.* **2006**, *27*, 3049–3055.

27. Veraverbeke, S.; Stavros, E.N.; Hook, S.J. Assessing fire severity using imaging spectroscopy data from the Airborne Visible/Infrared Imaging Spectrometer (AVIRIS) and comparison with multispectral capabilities. *Remote Sens. Environ.***2014**, *154*, 153–163.

28. Utkin, A.B.; Fernandes, A.; Simões, F.; Lavrov, A.; Vilar, R. Feasibility of forest-fire smoke detection using lidar. *Int. J. Wildland Fire* **2003**, *12*, 159–166.

29. Lavrov, A.; Utkin, A.B.; Vilar, R.; Fernandes, A. Evaluation of smoke dispersion from forest fire plumes using lidar experiments and modelling. *Int. J. Therm. Sci.* **2006**, *45*, 848–859.

30. Utkin, A.B.; Piedade, F.; Beixiga, V.; Mota, P.; Lousã, P. Scalable lidar technique for fire detection. *Proc. SPIE* **2014**,*9286*, 92860D–92860D-12.

31. Sahin, Y.G.; Ince, T. Early Forest Fire Detection Using Radio-Acoustic Sounding System. *Sensors* **2009**, *9*, 1485–1498.

32. Stipaničev, D. Intelligent Forest Fire Monitoring System—From idea to realization. In *Annual 2010/2011 of the Croatian Academy of Engineering*; Kluwer Academic Publishers: Hingham, MA, USA, 2012.

33. Surit, S.; Chatwiriya, W. Forest Fire Smoke Detection in Video Based on Digital Image Processing Approach with Static and Dynamic Characteristic Analysis. Proceedings of the 2011 First ACIS/JNU International Conference on Computers, Networks, Systems and

Industrial Engineering (CNSI), Jeju, Island, 23–25 May 2011; pp. 35–39.

34. Stula, M.; Krstinic, D.; Seric, L. Intelligent forest fire monitoring system. *Inf. Syst. Front.* **2012**, *14*, 725–739.

35. Owayjan, M.; Freiha, G.; Achkar, R.; Abdo, E.; Mallah, S. Firoxio: Forest fire detection and alerting system. Proceedings of the IEEE Mediterranean Electrotechnical Conference (MELECON), Beirut, Lebanon, 13–16 April 2014; pp. 177–181.

36. Mazzeo, G.; Filizzola, C.; Coviello, I.; Marchese, F.; Corrado, R.; Lacava, T.; Paciello, R.; Pergola, N.; Tramutoli, V. Performance assessment of a Robust Satellite Techniques (RST-FIRES) for forest fire detection and monitoring. Proceedings of the EGU General Assembly Conference, Vienna, Austria, 22–27 April 2012; Volume 14.

37. Milz, M.; Rymdteknik, A. *Study on Forest Fire Detection with Satellite Data*; Working Paper; Luleås Tekniska Universitet: LULEA, Sweden, 2013.

38. Schroeder, W.; Ruminski, M.; Csiszar, I.; Giglio, L.; Prins, E.; Schmidt, C.; Morisette, J. Validation analyses of an operational fire monitoring product: The Hazard Mapping System. *Int. J. Remote Sens.* **2008**, *29*, 6059–6066.

39. Phan, C.; Liu, H. A cooperative UAV/UGV platform for wildfire detection and fighting. Proceedings of the Asia Simulation Conference—7th International Conference on System Simulation and Scientific Computing, Beijing, China, 10–12 October 2008; pp. 494–498.

40. Ambrosia, V.G.; Wegener, S.; Zajkowski, T.; Sullivan, D.V.; Buechel, S.; Enomoto, F.; Lobitz, B.; Johan, S.; Brass, J.; Hinkley, E. The Ikhana unmanned airborne system (UAS) western states fire imaging missions: from concept to reality (2006–2010). *Geocarto Int.* **2011**, *26*, 85–101.

41. Merino, L.; Caballero, F.; Martínez-de Dios, J.R.; Maza, I.; Ollero, A. An Unmanned Aircraft System for Automatic Forest Fire Monitoring and Measurement. *J. Intell. Robot. Syst.* **2012**, *65*, 533–548.

42. Katzis, K.; Grondoudis, A.; Boustras, G.; Papazoglou, P. Forest fires: Proactive and reactive surveillance employing *in-situ*, aerial, and space technology. *Proc. SPIE* **2013**, *8795*, 87951K–87951K-11.

43. Navarro, M.; Davis, T.W.; Villalba, G.; Li, Y.; Zhong, X.; Erratt, N.; Liang, X.; Liang, Y. Towards Long-Term Multi-Hop WSN Deployments for Environmental Monitoring: An Experimental Network Evaluation. *J. Sens. Actuator Netw.* **2014**,*3*, 297–330.

44. Kosucu, B.; Irgan, K.; Kucuk, G.; Baydere, S. FireSenseTB: A Wireless Sensor Networks Testbed for Forest Fire Detection. Proceedings of the 2009 International Conference on Wireless Communications and Mobile Computing: Connecting the World Wirelessly, Leipzig, Germany, 21–24 June 2009; pp. 1173–1177.

45. Isaac, S.; Hancke, G.; Madhoo, H.; Khatri, A. A survey of wireless sensor network applications from a power utility's distribution perspective. Proceedings of the AFRICON, Livingstone, Zambia, 13–15 September 2011; pp. 1–5.

46. Rosi, A.; Berti, M.; Bicocchi, N.; Castelli, G.; Corsini, A.; Mamei, M.; Zambonelli, F. Landslide monitoring with sensor networks: Experiences and lessons learnt from a real-world deployment. *Int. J. Sens. Netw.* **2011**, *10*, 111–122.

47. Strazdins, G.; Elsts, A.; Nesenbergs, K.; Selavo, L. Wireless Sensor Network Operating System Design Rules Based on Real-World Deployment Survey. *J. Sens. Actuator Netw.* **2013**, *2*, 509–556.

48. Bouabdellah, K.; Noureddine, H.; Larbi, S. Using Wireless Sensor Networks for Reliable Forest Fires Detection.*Procedia Comput. Sci.* **2013**, *19*, 794–801.

49. Bahrepour, M.; Meratnia, N.; Havinga, P. *Automatic Fire Detection: A Survey from Wireless Sensor Network Perspective*; Technical Report TR-CTIT-08-73; University of Twente, Centre for Telematics and Information Technology: Enschede, The Netherlands, 2008.

50. Teguh, R.; Honma, T.; Usop, A.; Shin, H.; Igarashi, H. Detection and Verification of Potential Peat Fire Using Wireless Sensor Network and UAV. Proceedings of the International Conference Information Technology and Electrical Engineering, Yogyakarta, Indonesia, 12 July 2012; pp. 6–10.

51. Sanchez-Matamoros, J.; Dios, J.d.; Ollero, A. Cooperative localization and tracking with a camera-based WSN. Proceedings of the IEEE International Conference on Mechatronics, Malaga, Spain, 14–17 April 2009; pp. 1–6.

52. Aslan, Y.E.; Korpeoglu, I.; Ulusoy, Ö. A framework for use of wireless sensor networks in forest fire detection and monitoring. *Comput. Environ. Urban Syst.* **2012**, *36*, 614–625.

53. Kovács, Z.G.; Marosy, G.E.; Horváth, G. Case study of a simple, low power WSN implementation for forest monitoring. Proceedings of the 12th Biennial Baltic Electronics Conference (BEC), Tallinn, Estonia, 4–6 October 2010; pp. 161–164.

54. Bayo, A.; Antolin, D.; Medrano, N.; Calvo, B.; Celma, S. Development of a Wireless Sensor Network System for Early Forest Fire Detection. Proceedings of the 2010 European Workshop on Smart Objects: Systems, Technologies and Applications (RFID Sys Tech), Ciudad, Spain, 15–16 June 2010; pp. 1–7.

55. Lazarescu, M.T. Design of a WSN Platform for Long-Term Environmental Monitoring for IoT Applications. *IEEE J. Emerg. Sel. Top. Circuits Syst.* **2013**, *3*, 45–54.

56. Fierens, P.I. Number of wireless sensors needed to detect a wildfire. *Int. J. Wildland Fire* **2009**, *18*, 625–629.

57. Bogue, R. Sensors for fire detection. *Sens. Rev.* **2013**, *33*, 99–103.

58. Fernández-Berni, J.; Carmona-Galán, R.; Leñero-Bardallo, J.A.; Kleihorst, R.; Rodríguez-Vázquez, Á. *Towards an Ultra-Low-Power Low-Cost Wireless Visual Sensor Node for Fine-Grain Detection of Forest Fires*; Imprensa da Universidade de Coimbra: Coimbra, Portugal, 2014.

59. Lloret, J.; Garcia, M.; Bri, D.; Sendra, S. A wireless sensor network deployment for rural and forest fire detection and verification. *Sensors* **2009**, *9*, 8722–8747.

60. Cheong, P.; Chang, K.F.; Lai, Y.H.; Ho, S.K.; Sou, I.K.; Tam, K.W. A ZigBee-Based Wireless Sensor Network Node for Ultraviolet Detection of Flame. *IEEE Trans. Ind. Electron.* **2011**, *58*, 5271–5277.

61. Manolakos, E.; Logaras, E.; Paschos, F. Wireless Sensor Network Application for Fire Hazard Detection and Monitoring. In *Sensor*

Applications, Experimentation, and Logistics; Lecture Notes of the Institute for Computer Sciences, Social Informatics and Telecommunications Engineering; Komninos, N., Ed.; Springer: Berlin/Heidelberg, Germany, 2010; Volume 29, pp. 1–15.

62. Shixing, L.; Defeng, T.; Yongming, Z. Multiparameter fire detection based on wireless sensor network. Proceedings of the IEEE International Conference on Intelligent Computing and Intelligent Systems, Shanghai, China, 20–22 November 2009; Volume 3, pp. 203–206.

63. Zervas, E.; Mpimpoudis, A.; Anagnostopoulos, C.; Sekkas, O.; Hadjiefthymiades, S. Multisensor data fusion for fire detection. *Inf. Fusion* **2011**, *12*, 150–159.

64. Sathik, M.M.; Mohamed, M.S.; Balasubramanian, A. Fire detection using support vector machine in wireless sensor network and rescue using pervasive devices. *Int. J. Adv. Netw. Appl.* **2010**, *2*, 636–639.

65. Soliman, H.; Sudan, K.; Mishra, A. A smart forest-fire early detection sensory system: Another approach of utilizing wireless sensor and neural networks. Proceedings of IEEE Sensors Conference, Kona, HI, USA, 1–4 November 2010; pp. 1900–1904.

66. Bahrepour, M.; Meratnia, N.; Havinga, P. Sensor fusion-based event detection in Wireless Sensor Networks. Proceedings of the Annual International Mobile and Ubiquitous Systems: Networking Services, Toronto, ON, Canada, 13–16 July 2009; pp. 1–8.

67. Somov, A.; Spirjakin, D.; Ivanov, M.; Khromushin, I.; Passerone, R.; Baranov, A.; Savkin, A. Combustible Gases and Early Fire Detection: An Autonomous System for Wireless Sensor Networks. Proceedings of the 1st International Conference on Energy-Efficient Computing and Networking, Passau, Germany, 13–15 April 2010; pp. 85–93.

68. Hartung, C.; Han, R.; Seielstad, C.; Holbrook, S. FireWxNet: A Multi-tiered Portable Wireless System for Monitoring Weather Conditions in Wildland Fire Environments. Proceedings of the 4th International Conference on Mobile Systems, Applications and Services, Uppsala, Sweden, 19–22 June 2006; pp. 28–41.

69. Bapat, S.; Kulathumani, V.; Arora, A. Analyzing the yield of ExScal, a large-scale wireless sensor network experiment. Proceedings of the

IEEE International Conference on Network Protocols, Boston, MA, USA, 6–9 November 2005; p. 10.

70. Dutta, P.; Hui, J.; Jeong, J.; Kim, S.; Sharp, C.; Taneja, J.; Tolle, G.; Whitehouse, K.; Culler, D. Trio: Enabling Sustainable and Scalable Outdoor Wireless Sensor Network Deployments. Proceedings of the 5th International Conference on Information Processing in Sensor Networks, Nashville, TN, USA, 19–21 April 2006; pp. 407–415.

71. Wark, T.; Hu, W.; Corke, P.; Hodge, J.; Keto, A.; Mackey, B.; Foley, G.; Sikka, P.; Brunig, M. Springbrook: Challenges in developing a long-term, rainforest wireless sensor network. Proceedings of the International Conference on Intelligent Sensors, Sensor Networks and Information Processing, Sydney, NSW, Australia, 15–18 December 2008; pp. 599–604.

72. Neubauer, B.; Sidén, J.; Olofsson, C.; Gulliksson, M.; Koptyug, A.; Nilsson, H.E.; Norgren, M. A new thermally activated battery cellbased forest fire detection and monitoring system. *Model. Monit. Manag. For. Fires III* **2012**, *158*.

73. Tu, Y.F. *Assessment of the Current False Alarm Situation from Fire Detection Systems in New Zealand and the Development of an Expert System for Their Identifications*; Research Report 02/10; University of Canterbury, Civil Engineering: Christchurch, New Zealand, 2002.

74. Díaz-Ramírez, A.; Tafoya, L.A.; Atempa, J.A.; Mejía-Alvarez, P. Wireless Sensor Networks and Fusion Information Methods for Forest Fire Detection. *Procedia Technol.* **2012**, *3*, 69–79.

75. Sudevalayam, S.; Kulkarni, P. Energy Harvesting Sensor Nodes: Survey and Implications. *Commun. Surv. Tutor.* **2011**,*13*, 443–461.

76. Bandyopadhyay, S.; Chandrakasan, A. Platform architecture for solar, thermal and vibration energy combining with MPPT and single inductor. Proceedings of the VLSI Circuits, VLSIC, VLSIC, Honolulu, HI, USA, 15–17 June 2011; pp. 238–239.

77. Cano, C.; Bellalta, B.; Sfairopoulou, A.; Barcelo, J. A low power listening MAC with scheduled wake up after transmissions for WSNs. *Commun. Lett.* **2009**, *13*, 221–223.

78. El-Hoiydi, A. Aloha with preamble sampling for sporadic traffic in ad hoc wireless sensor networks. *IEEE Int. Conf. Commun.* **2002**, *5*, 3418–3423.

79. Watteyne, T.; Vilajosana, X.; Kerkez, B.; Chraim, F.; Weekly, K.; Wang, Q.; Glaser, S.; Pister, K. OpenWSN: A standards-based low-power wireless development environment. *Trans. Emerg. Telecommun. Technol.* **2012**, *23*, 480–493.

80. Kuorilehto, M.; Kohvakka, M.; Suhonen, J.; Hämäläinen, P.; Hännikäinen, M.; Hämäläinen, T.D. *Ultra-Low Energy Wireless Sensor Networks in Practice*; John Wiley & Sons, Ltd: New York, NY, USA, 2007.

81. Liu, Y.; He, Y.; Li, M.; Wang, J.; Liu, K.; Mo, L.; Dong, W.; Yang, Z.; Xi, M.; Zhao, J.; Li, X.Y. Does wireless sensor network scale? A measurement study on GreenOrbs. In. Proceedings of the IEEE INFOCOM, Shanghai, China, 10–15 April 2011; pp. 873–881.

82. Rahman, F.; Devanbu, P. How, and Why, Process Metrics Are Better. Proceedings of the 2013 International Conference on Software Engineering, San Francisco, CA, USA, 18–26 May 2013; pp. 432–441.

83. Sun, D.; Lu, J.; Wang, W.; Lin, Z. Design of WSN nodes and network performance analysis in a tea plantation. Proceedings of IET International Conference on Wireless Sensor Network, Beijing, China, 15–17 November 2010; pp. 144–147.

84. Micek, J.; Kapitulik, J. WSN sensor node for protected area monitoring. Proceedings of 2012 Federated Conference on Computer Science and Information Systems (FedCSIS), Wroclaw, Poland, 9–12 September 2012; pp. 803–807.

85. Flickenger, R.; Okay, S.; Pietrosemoli, E.; Zennaro, M.; Fonda, C. Very long distance Wi-Fi networks. Proceedings of the Second ACM SIGCOMM Workshop on Networked Systems for Developing Regions, Seattle, WA, USA, 17–22 August 2008; pp. 1–6.

86. Celandroni, N.; Ferro, E.; Gotta, A.; Oligeri, G.; Roseti, C.; Luglio, M.; Bisio, I.; Cello, M.; Davoli, F.; Panagopoulos, A.D.; *et al.* A survey of architectures and scenarios in satellite-based wireless sensor networks: System design aspects.*Int. J. Satell. Commun. Netw.* **2013**, *31*, 1–38.

87. Raman, B.; Chebrolu, K. Censor networks: A critique of "sensor networks" from a systems perspective. *SIGCOMM Comput. Commun. Rev.* **2008**, *38*, 75–78.

88. Boan, J. Radio Experiments With Fire. *IEEE Antennas Wirel. Propag. Lett.* **2007**, *6*, 411–414.

89. Figueiredo, C.; Nakamura, E.; Ribas, A.; de Souza, T.; Barreto, R. Assessing the communication performance of wireless sensor networks in rainforests. Proceedings of 2009 2nd IFIP Wireless Days (WD), Paris, France, 15–17 December 2009; pp. 1–6.

90. SystemC. Available online: http://www.systemc.org/ (accessed on 17 April 2015).

91. Reed, C.; Botts, M.; Davidson, J.; Percivall, G. OGC® sensor web enablement: overview and high level achhitecture. Proceedings of the IEEE Autotestcon, Baltimore, MD, USA, 17–20 September 2007; pp. 372–380.

92. Botts, M.; Percivall, G.; Reed, C.; Davidson, J. OGC® Sensor Web Enablement: Overview and High Level Architecture. In *GeoSensor Networks*; Lecture Notes in Computer Science; Nittel, S., Labrinidis, A., Stefanidis, A., Eds.; Springer: Berlin Heidelberg, Germany, 2008; Volume 4540, pp. 175–190.

93. Jirka, S.; Bröring, A.; Stasch, C. Discovery Mechanisms for the Sensor Web. *Sensors* **2009**, *9*, 2661–2681.

94. Rouached, M.; Baccar, S.; Abid, M. RESTful Sensor Web Enablement Services for Wireless Sensor Networks. In. Proceedings of the IEEE Eighth World Congress on Services, Honolulu, HI, USA, 24–29 June 2012; pp. 65–72.

7

AN INTELLIGENT WIRELESS SENSOR NETWORK TEMPERATURE ACQUISITION SYSTEM WITH AN FPGA

Rachid Souissi[1], Mohsen Ben-Ammar[2]

[1]Computer Science Department, Higher Technological Institute of Sfax (ISET), Sfax, Tunisia

[2]Electrical Engineering Department, National Engineering School of Sfax (ENIS) Machine Control and Power Grid Research Unit, CMERP, Sfax, Tunisia

ABSTRACT

The use of FPGAs in the design of wireless sensor networks can improve the processing system performance so as to develop sensor nodes with powerful embedded processor. In fact, FPGA has a super-parallel processing capacity in data processing and its use and configuration is flexible. This allows intensive digital signal processing tasks like Fast Fourier Transforms to be implemented on the sensor node and this can even support intelligent applications and new security algorithms. This paper presents a wireless sensor network system that monitors and visualizes periodically the environment temperature. The proposed system consists of a base station node based on an FPGA DE2-70 and five sensor nodes having the architecture of the System-on-Chip CC2430.

KEYWORDS

Wireless Sensor Network; ZigBee; System-on-Chip CC2430; FPGA DE2-70

1. INTRODUCTION

The use of wireless sensor networks (WSNs) has risen in last years. In fact, several industrials, researchers and engineers work on this new technology both in the civilian and military sectors that require the use of many nodes, even hundreds or thousands. That is why more applications have been developed in order to solve several problems in data acquisition and environment control. Moreover, these nodes must be capable of sensing, processing and communicating physical parameters like temperature and pressure through the global wireless network. This paper focuses on the design of a WSN to acquire the temperature from some autonomous nodes. It introduces the use of the FPGA DE2-70 in this application field. The use of the FPGA DE2-70 in WSN can improve the processing system performance so as to develop powerful embedded processor. This paper highlights the different techniques, methods and protocols that can be used in this wireless sensor network technology. The developed system permits monitoring the environment temperature.

The organization of this paper is as follows; In Section 2, we will present some works dealing with sensor nodes mainly based on an FPGA, otherwise Section 3 deals with the IEEE 802.15.4 standard that defines the physical layer in all ZigBee devices; this standard is used by the sensor nodes to communicate with each other. Section 4 is devoted to the system hardware architecture; it exposes the main components of the system. The software architecture is exposed in Section 5. The application description is presented in Section 6. Finally, a conclusion is given to the whole work.

2. RELATED WORK

In this section we will present an in-depth knowledge of several wireless sensor networks systems based on FPGA. The reference paper

[1] provides an interface and a software/hardware module for Intelligent Car Park Management System based on FPGA. So, in car entering module, as the car enters the lot, it is detected by the IR sensors. The IR sensors provide the pulse to the FPGA which assumes that an input is detected and thus the car is entered into the parking lot. As the car leaves the lot, it is detected by the IR sensors. The IR sensors provide the pulse to the FPGA which assumes that an input is detected and thus the car is exited out of the parking lot.

The project [2] has been carried out to develop a novel FPGA-based photosynthesis smart sensor. The smart sensor is capable of acquiring and fusing the primary sensor signals to measure temperature, relative humidity, solar radiation, CO_2, air pressure and air flow. This permits the estimation of carbon balance, integrative and derivative variables from net photosynthesis in real time due to the FPGA processing capabilities. To accomplish this, high performance computational capabilities are required. These days, Field Programmable Gate Arrays (FPGAs) are devices that are being employed in applications where high demand computational resources are necessary (Contreras-Medina et al., 2012). Due to the demands for high computation required in this research the use of FPGA-based smart sensor with capabilities of acquiring the primary sensor readings, memory management, signal processing, photosynthesis calculation, wireless communication management and in situ real time photosynthesis signal visualization in a color LCD screen is proposed.

In this work [3], we have developed a modular node. So, both digital and analogue type sensors are being considered. The signals that they produce are conditioned for interfacing to the processing and communications modules. For processing there is a field programmable gate array (FPGA) module. This allows intensive digital signal processing (DSP) tasks such as moving average filters, Fast Fourier Transforms (FFTs) to be implemented on the sensor node and this can even support intelligent applications. The FPGA module was included as this provides a reconfigurable processing solution in which intensive DSP Tasks or in which intelligent functionality can be implemented.

In [4], Wei et al. presented a sensor node based on SoC. Such SoC is based on the OpenRisc1200 processor and executes the μ/OS II RTOS and Wishbone bus. The SoC consists of three main components: an acquisition unit that is fed from a digital thermometer coupled to the FPGA, the OpenRisc firmware that processes the temperature and a logic block that interfaces a ZigBee transceiver based on RS-232 (SHUNCOM SZ05-STD). The utilized FPGA is an Altera Cyclone II

(EP2C70). The system consumes 221 mW. Nevertheless, given that the FPGA only interfaces several components (digital thermometer and ZigBee transmitter), it seems to be an overkill, since its inner embedded resources are not used and the same task can be performed using a low-power microcontroller.

3. THE COMMUNICATION STACK

The proposed system uses the IEEE 802.15.4 standard [5] as the communication protocol. The IEEE 802.15.4 standard defines the characteristics of the physical and MAC layers for Low-Rate Wireless Personal Area Networks (LR-WPAN) including wireless sensor networks (WSNs). IEEE 802.15.4 focuses mainly on low-cost, low-power communication between devices and therefore it presents a low transfer rate with a maximum of 250 kbits/s.

3.1. Physical (PHY) Layer

The IEEE 802.15.4 standard defines the physical layer (PHY) in all ZigBee devices. The PHY is responsible for data transmission and reception by using a defined radio channel and specific modulation and spreading technique [6]. The IEEE 802.15.4 standard specifies two physical layers that represent three operational frequency bands. These three bands include: 868 MHz (used in Europe), 915 MHz (used in America), and 2.4 GHz (used worldwide) [7]. The 868 and 915 MHz bands are in one PHY using the Binary Phase Shift Keying (BPSK) for modulation, while the 2.4 GHz band is in the second PHY and employs Offset Quadrature Phase Shift Keying (O-QPSK) for modulation. There is a single channel between 868 and 868.8 MHz, 10 channels between 902 and 928 MHz, and 16 Channels between 2.4 and 2.4835 GHz, all using the Direct Sequence Spread Spectrum (DSSS) access mode[6] .

3.2. Medium Access Control (MAC) Layer

In addition to the physical layer, the IEEE 802.15.4 standard defines the medium access control layer for all ZigBee devices. The MAC layer protocol serves as the interface between the PHY and the higher layer protocols. The functions of the MAC include synchronization, frame validation, acknowledged frame delivery, association, and disassociation [5,8].

The MAC layer defines two types of devices; Full Function Device (FFD) and Reduced Function Device (RFD).

A FFD node implements a full set of MAC layer functions and can communicate with both FFD as well as RFD devices. However, an RFD device implements only a reduced set of functions and it is necessary attached to an FFD node playing the role of a network coordinator (called also PAN coordinator). The PAN coordinator select one of the two operational communication modes defined in IEEE 802.15.4, namely, the non beacon-enabled mode and the beacon-enabled mode, in which beacons are periodically sent by the PAN coordinator to identify and synchronize communication inside its PAN. In both modes, the MAC controls the access to the radio channel by employing some methods like the Carrier Sense Multiple Access with Collision Avoidance (CSMA/CA) mechanism. CSMA/CA is a network contention protocol that listens to the network before sending, in order to avoid collision [8].

3.3. ZigBee Layers

Based on the IEEE 802.15.4, the ZigBee standard [9] defines the higher layer namely; the network layer and the application layer (**Figure 1**).

The network layer is responsible for joining/leaving a network, security, routing, discovering 1-hop neighbors and storing neighbor information. The ZigBee network layer supports three topologies; the star topology where end devices (RFD) are attached to a central point playing the role of PAN coordinator (FFD), the tree topology where end devices can be attached also to FFD nodes with routing capabilities playing the role of ZigBee routers in a hierarchical manner (with parent-child relationship), the mesh topology where ZigBee routers can be fully connected.

As shown in **Figure 1** , the application layer includes the Application Framework, the ZigBee Device Objects (ZDO), and the Application Sub Layer (APS). The Application Framework can have up to 240 Application Objects (APOs), that is, user defined application modules which are part of a ZigBee application [6]. The ZDO defines the role of the device, initiates and responds to binding requests and establishes a secure relationship between devices. The APS offers an interface to data and security services to the APO sand ZDO.

4. HARDWARE ARCHITECTURE

As shown in **Figure 2**, our system consists of three types of entities. An end device is a sensor nodes composed by one System-on-Chip (CC2430) and is responsible for sensing environment parameters (such as temperature) and then sending the sensed data to the base station. The base station contains a System-on-Chip (CC2430) and is connected to an FPGA DE2-70 through its expansion header. The base station is responsible for collecting data from end devices. After that, the FPGA DE2-70 reads this information then sends it joined with the address of the appropriate node to the personal computer through one serial link. The PC is responsible for visualizing the received information for further analysis and interpretation [10].

4.1. The FPGA DE2-70

The Altera FPGA is widely used as the programmable device to support the Design and Reuse (D & R). This is even more convenient with the development of System on Programmable Chip (SOPC) where Altera is leading the industry with its FPGA chips and the Quartus Application tools. The Altera Nios is 32-bit softcore RISC processor that is optimized for Altera Field Programmable Gate Array (FPGA) devices. It is a powerful embedded processor and is mainly used to control reconfigurable circuits running on the Altera FPGA. A schema of the DE2-70 board is shown in **Figure 3**. It depicts the layout of the board and indicates the location of the connectors and key components.

4.2. BB—Battery Board

This section describes the System-on-Chip battery board (BB). The main function for this board is to power the CC2430EM (Evaluation Module) or CC2431EM with the use of two AA batteries. It can also be powered by a lab powered connected directly to GND and VDD on the board. The Low-Power RF BB is a simple battery module for use with an EM. As presented in **Figure 4**, it has one LED, one push switch, one power switch and I/O connector A and B gives access to all I/O on the SoC and to some additional pins. It allows you quickly deploy a SOC network [11]. The photograph of the component is shown in the **Figure 4**. On this Battery Board, we can find the Evaluation Module, which contains the minimum Components for a RF part to function.

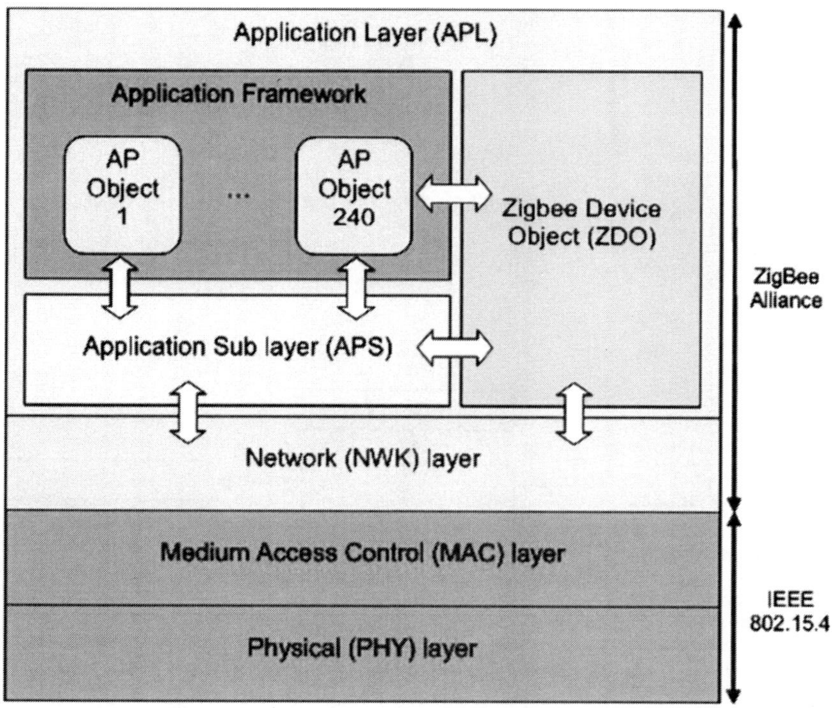

Figure 1. The communication stack.

4.3. System-on-Chip CC2430

The CC2430 comes in three different versions: CC2430- F32/64/128, with 32/64/128 KB of flash memory respectively. The CC2430 is a true System-on-Chip (SoC) solution specifically tailored for IEEE 802.15.4 and ZigBee™ applications. It enables ZigBee™ nodes to be built with very low total bill-of material costs.

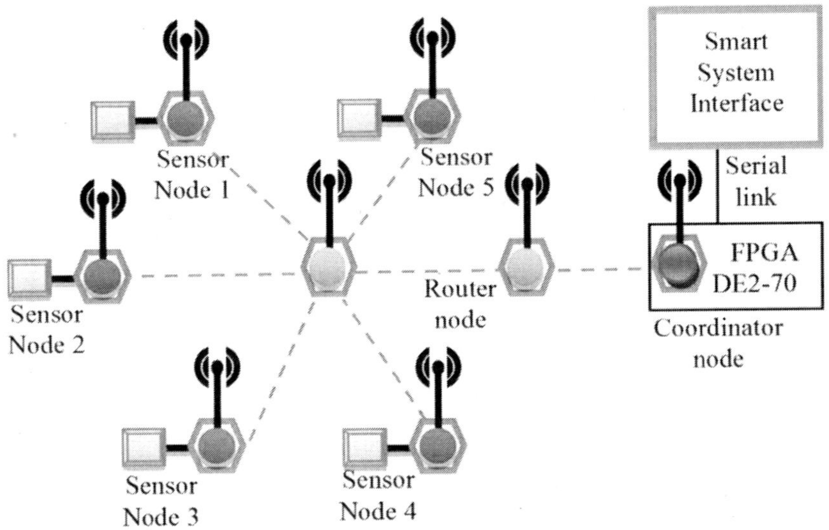

Figure 2. System architecture.

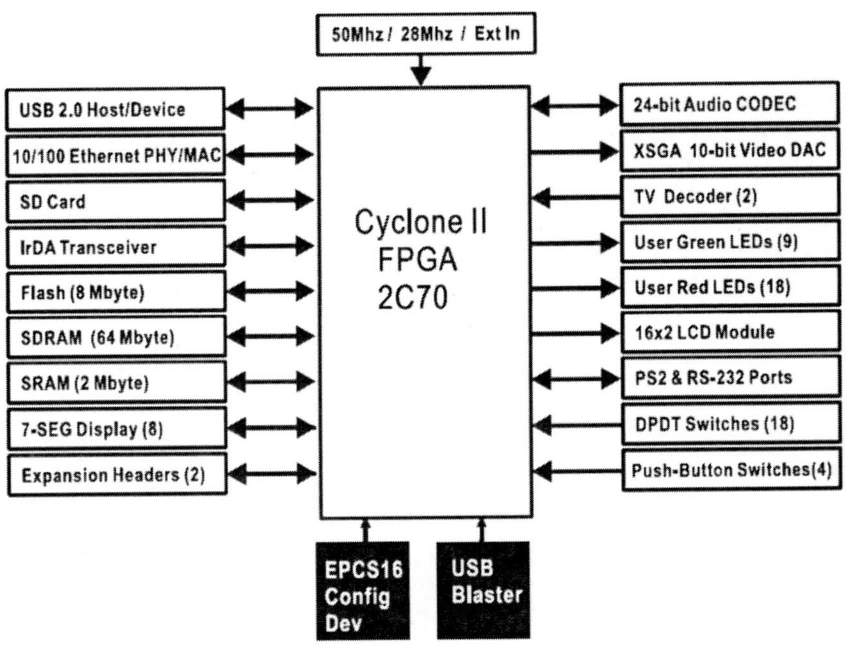

Figure 3. FPGA DE2-70 Board.

Figure 4. System-on-Chip Battery Board with CC2430EM.

The CC2430 combines the excellent performance of the leading CC2420 RF transceiver with an industrystandard enhanced 8051 MCU, 32/64/128 KB flash memory, 8 KB RAM and many other powerful features. Combined with the industry leading ZigBee™ protocol stack (Z-Stack), the CC2430 provides the market's most competitive ZigBee™ solution.

The CC2430 is highly suited for systems where ultralow power consumption is required. This is ensured by various operating modes. Short transition times between operating modes further ensure low power consumption.

4.4. 8051 CPU

The CC2430 includes an 8-bit CPU core which is an enhanced version of the industry standard 8051 core. The enhanced 8051 core uses the standard 8051 instruction set. Instructions execute faster than the standard 8051 due to the following:

- One clock per instruction cycle is used as opposed to 12 clocks per instruction cycle in the standard 8051;

- Wasted bus states are eliminated. Since an instruction cycle is aligned with memory fetch when possible, most of the single byte instructions are performed in a single clock cycle.

5. SOFTWARE ARCHITECTURE

The design of the software in this package is based on the layered architecture as depicted in **Figure 5**.

Application layer: This Software package contains several applications examples with access to Basic RF and HAL (Hardware Abstraction Layer) [12].

Basic RF: This layer offers a simple protocol for transmission and reception on two-way RF link.

Hardware Abstraction Layer: Contains functionality for access to the radio and on-board peripherals modules like LCD, UART, joysticks, buttons, and timers [5].

5.1. Basic RF

The Basic RF layer offers a simple protocol for transmission and reception on a two-way RF link. The Basic RF protocol offers the service for packet transmission and reception. It also offers secure communication by use of CCM-64 authentication and encryption/ decryption of packets. The security features of Basic RF can be compiled in by defining the compile switch SECURITY_ CCM in the project file. The compile time inclusion of security features is done to save code space for the applications where security features are not needed [5].

The protocol uses IEEE 802.15.4 MAC compliant data and acknowledgment packets. However, it does not offer a full MAC layer, only a simple data link layer for communication between two nodes. Basic RF contains only a small subset of the 802.15.4 standard:

- Association, scanning or beacons are not implemented;
- No defined coordinator/device roles (peer-to-peer, all nodes are equal);

- No packet retransmission. This must be taken care of by the layer above Basic RF.

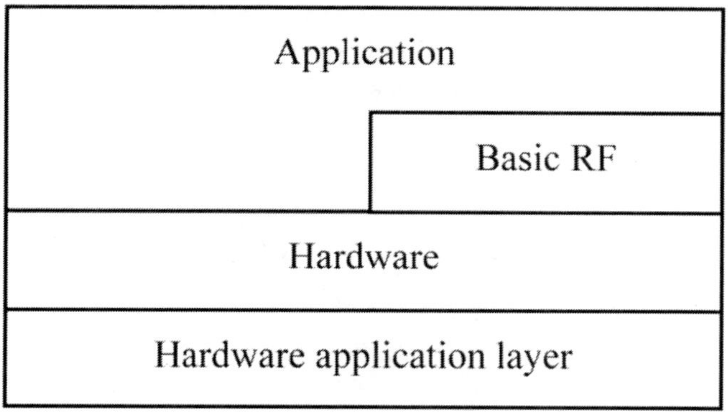

Figure 5. Software architecture.

5.2. Basic RF Instructions

5.2.1. Startup

1) Make sure that the board peripherals and radio interface is initialized i.e. halBoardInit() must have been called first.

2) Create a basicRfCfg_t structure, and initialize its members. If the security features of Basic RF are used, the higher layer is responsible for allocating and assigning the 16 bytes key.

3) Call basicRfInit() to initialize the packet protocol.

5.2.2. Transmission

1) Create a buffer with the payload to send. Maximum payload size for Basic RF is 103 Bytes.

2) Call basicRfSendPacket(). Check the return value.

5.2.3. Reception

1) Perform polling by calling basicRfPacketIsReady() to check if a new packet is ready to be received by the higher layer.

2) Call basicRfReceive() to receive the packet by higher layer. The caller is responsible for allocating a buffer large enough for the packet and 2 Bytes buffer space for the RSSI value [13].

5.3. IAR Embedded Workbench� for 8051

IAR Embedded Workbench is a set of highly sophisticated and easy-to-use development tools for embedded applications. It integrates the IAR C/C++ Compiler™, assembler, linker, librarian, text editor, project manager, and C-SPY� Debugger in an integrated development environment (IDE). With its built-in chip-specific code optimizer, IAR Embedded Workbench generates very efficient and reliable FLASH/PROMable code for the 8051 microcontroller. In addition to this solid technology, IAR Systems also provides professional worldwide technical support. So, we have used this environment to write then download program of each node inside his FLASH EPROM.

6. RESULTS AND DISCUSSION

Our application is composed by two modules "Node()" and "FPGADE270()". The "Node()" moduleis executed on the end devicenodes, while the "FPGADE270()" module is executed on the base stationnode $CC2430_{(bs)}$ linked to the FPGA DE2-70. The module "Node()" will read the temperature of each node and send the appropriate value joined with the address of the node to the main node $CC2430_{(bs)}$ linked to the FPGA DE2-70. When the information is received, it will be sent to the Personal Computer through the serial port where it will be visualised.

6.1. The "Node" Module

In this module we have implemented a function that permits to read the temperature. This function will read the temperature of the node, then return the appropriate value to the parent module executed on the node. In fact, in order to get the temperature we have used the predefined functionadcSampleSingle(ADC_REF_1_25_V,ADC_12_ BIT, ADC_TEMP_SENS).

```
#include <adc.h>
#ifndef _HAL_TEMP_H
#define _HAL_TEMP_H
#define ADC_REF_1_25_V0x00// Internal 1.25V reference
#define ADC_12_BIT 0x30 // 512 decimation rate
#define ADC_TEMP_SENS 0x0E// on-chip temperature sen-
sor
#define CONST 0.61065 // (1250 / 2047)
#define OFFSET_DATASHEET 743
#define
OFFSET_MEASURED_AT_25_DEGREES_CELCIUS
29.75
#define OFFSET (OFFSET_DATASHEET +
OFFSET_MEASURED_AT_25_DEGREES_CELCIUS)
// 779.75
#define TEMP_COEFF 2.45
unsigned char getTemp(void)
    {
  unsignedintadcValue;
  floatoutputVoltage;
  ad-
  cValue=adcSampleSingle(ADC_REF_1_25_V,AD
  C_12_BIT,ADC_TEMP_SENS);
  adcValue>>= 4;
  outputVoltage = adcValue * CONST;
  return ((outputVoltage - OFFSET) /
  TEMP_COEFF); }
```

When the temperature is acquired, the node will send it to the main node or to his neighbour in general, so we have developed the following module in order to do this RF transmission. So, in the following module we will call the function getTemp() to get the temperature of each node.

```
#define PAN_ID 0x2007
#define CLIENT_ADDR 0x2524
#define SERVEUR_ADDR 0xBEEF
#define APP_PAYLOAD_LENGTH 3
static uint8 pTxData[APP_PAYLOAD_LENGTH];

#define a 0
#define b 1
static void Node()
{
pTxData[a] = CLIENT_ADDR;
basicRfConfig.myAddr = SERVEUR_ADDR;
if(basicRfInit(&basicRfConfig)==FAILED)
{   HAL_ASSERT(FALSE); }
while (TRUE) {
halLedToggle(1);
        pTxData[b] = getTemp();
        basicRfSend-
Packet(SERVEUR_ADDR,pTxData,APP_PAYLOAD_
LENGTH);
        halIntOff();
        halMcuSetLowPowerMode(HAL_MCU_LPM_3);
        halIntOn();
}
}
```

6.2. The FPGADE270 Module

```
#define APP_PAYLOAD_LENGTH              3
#define a 0
#define b 1
static uint8 pRxData[APP_PAYLOAD_LENGTH];
static void FPGADE270()
{     while (TRUE) {
      while(!basicRfPacketIsReady());
basicRfReceive(pRxData,   APP_PAYLOAD_LENGTH,
NULL);
      long temp=pRxData[b];
      P0DIR=0xFF;
      P0=pRxData[a];
      halMcuWaitMs(100);
      P0=pRxData[b];
}}
```

In this module, we receive the appropriate value of the temperature of the node. In fact, to reach this goal we have used the predefined function basicRfReceive(pRx Data, APP_PAYLOAD_LENGTH, NULL) to read the information send by each node. This information is stored in the vector pRxData. So, the temperature is stored in pRxData[b] while the address of the node is in the pRxData[a].

6.3. Monitoring Interface

When the temperature is received by the FPGADE270 node it will be send later through the serial port to the personal computer where it will be visualised by one interface developed with Matlab Version R2012a as depicted in the following figures. So, the **Figure 6**(a) illustrates the results taken in hot period like June. Otherwise, in the **Figure 6**(b) we can see the results related to cold period like December month.

In this interface we can see an important increase of temperature values send by each node and other information like the time of each received information.

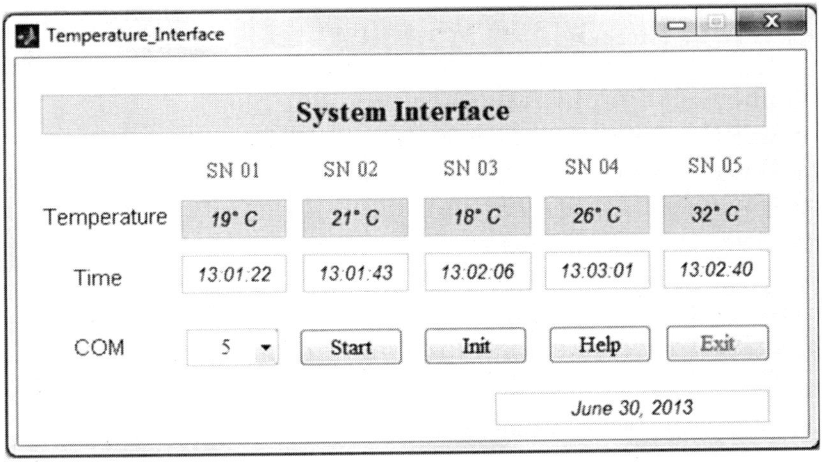

(a)

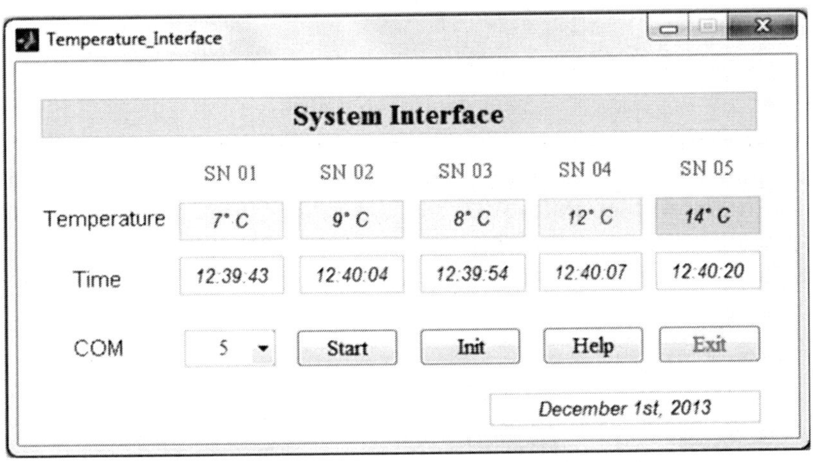

(b)

Figure 6. (a) Acquisition results in June the 30th 2013; (b) Acquisition results in December the 1st 2013.

7. CONCLUSIONS

The development of wireless sensor network reveals new challenges to engineers in several fields. In this context, the development of systems based on wireless sensor networks becomes a necessity. Therefore, actually we are obliged to develop some systems based on this new technology to solve several problems in data acquisition and control in different environments.

Our system includes two aspects: hardware and software. The hardware is composed of one base node $CC2430_{(bs)}$ connected to the FPGA DE2-70 and five nodes $CC2430_{(1..5)}$. Each component contains a Systemon-Chip CC2430 that contains the processor which executes the necessary program on each node. The application permits monitoring temperature of the environment. The future work is to develop a synchronisation algorithm to coordinate between transmissions of each node. Moreover, the information received by the WSN will be visualised through the internet network in order to benefit from the distant control and monitoring.

In our system, we have used five nodes to acquire the temperature and send it to the FPGA DE2-70 as a final destination, but if we want to cover a large distance, we can use more than five nodes, for example twenty nodes and the distance between two nodes is twenty meter, so each node can send his information to his neighbour node until the final destination or the main node $CC2430_{(bs)}$. So, the role of each node is to acquire the temperature from his location, receive and send information to the next node until we reach the last destination. By this way, the WSN will be installed in large area, in order to supervise it.

REFERENCES

1. R. Khan, Y. Ali Shah, Z. Khan, K. A. Muhammad and A. M. Amjad Ali, "Intelligent Car Parking Management System on FPGA," IJCSI International Journal of Computer Science Issues, Vol. 10, No. 1, 2013, pp. 171-173.

2. J. R. Millan-Almaraz, I. Torres-Pacheco, C. Duarte-Galvan, R. G. Guevara-Gonzalez, L. Miguel Contreras-Medina, R. de Jesus Romero-Troncoso and J. R. Rivera-Guillen, "FPGA-Based Wireless Smart Sensor for Real-Time Photosynthesis," Computers and Electronics in Agriculture, Vol. 93, 2013, pp. 58-69. www.elsevier.com/ locate/ compag

3. S. J. Bellis, K. Delaney, B. O'Flynn, J. Barton, K. M. Razeeb and C. O'Mathuna, "Development of Field Programmable Modular Wireless

Sensor Network Nodes for Ambient Systems," University College Cork, Cork, 2005.

4. J. Wei, L. Wang, F. Wu, Y. Chen and L. Ju, "Design and Implementation of Wireless Sensor Node Based on Open Core," Proceedings of the IEEE Youth Conference on Information, Computing and Telecommunication, Beijing, China, 20-21 September 2009, pp. 102-105.

5. Institute of Electrical and Electronics Engineers, Inc., IEEE Std.802.15.4-2003, "Wireless Medium Access Control (MAC) and Physical Layer (PHY) Specifications for Low Rate Wireless Personal Area Networks (LRWPANs)," IEEE Press, New York, 2003.

6. P. Baronti, P. Pillai, V. WC. Chook, S. Chessa, A. Gotta and Y. Fun Hu, "Wireless Sensor Networks: A Survey on the State of the Art and the 802.15.4 and ZigBee Standards," Computer Communications, Vol. 30, No. 7, 2007, pp. 1655-1695.

7. K. Scott, "Design and Performance of IEEE 802.15.4. Compliant MSE Receivers," Asilomar Conference on Signals, Systems and Computers, Vol. 2, November 2004, pp. 2051-2055.

8. G. Ding, B. Bhargava, Z. Sahinoglu and P. Zhang, "Reliable Broadcast in ZigBee Networks," IEEE SECON Proceedings, 2005, pp. 510-520.

9. ZigBee Alliance, "ZigBee Specifications," Version 1.0, 2005.

10. CC2430 Software Examples Users Guide TEXAS INSTRUMENTS.

11. Chipcon Products from Texas Instruments, CC2431DK Development Kit User Manual Rev. 1.5.

12. Y. E. Krasteva, J. Portilla, M. Carnicer, E. de la Torre and T. Riesgo, "Remote HW-SW Reconfigurable Wireless Sensor Nodes," Industrial Electronics, 2008.

13. J. Portilla, A. de Castro, E. de la Torre and T. Riesgo, "A Modular Architecture for Nodes in Wireless Sensor Networks," Journal of Universal Computer Science, Vol. 12, No. 3, 2006, pp. 328-339.

8

ON THE ROUTING PROTOCOL INFLUENCE ON THE RESILIENCE OF WIRELESS SENSOR NETWORKS TO JAMMING ATTACKS

Carolina Del-Valle-Soto [1,*], Carlos Mex-Perera [1], Raul Monroy [2] and Juan Arturo Nolazco-Flores [1]

[1] Department of Electrical and Computer Engineering, Tecnológico de Monterrey, Ave. Eugenio Garza Sada #2501 Sur, Monterrey, NL 64849, Mexico

[2] School of Science and Engineering, Tecnológico de Monterrey, Campus Estado de México, Carretera al lago de Guadalupe Km 3.5, Col. Margarita M. de Juárez, Atizapán 52926, Mexico

ABSTRACT

In this work, we compare a recently proposed routing protocol, the multi-parent hierarchical (MPH) protocol, with two well-known protocols, the *ad hoc* on-demand distance vector (AODV) and dynamic source routing (DSR). For this purpose, we have developed a simulator, which faithfully reifies the workings of a given protocol, considering a fixed, reconfigurable *ad hoc* network given by the number and location of participants, and general network conditions. We consider a scenario that can be found in a large number of wireless sensor network applications, a single sink node that collects all of the information generated by the sensors. The metrics used to compare the protocols were the number of packet retransmissions, carrier sense multiple access (CSMA) inner loop retries, the number of nodes answering the queries from the coordinator (sink) node and the energy consumption. We tested the network under ordinary (without attacks) conditions (and combinations thereof) and when it is subject to different types of jamming attacks (in particular, random and reactive jamming attacks), considering several positions for the jammer. Our results report that MPH has a greater ability to tolerate such attacks than DSR and AODV, since it minimizes and encapsulates the network segment under attack. The self-configuring capabilities of MPH derived from a combination of a proactive routes update, on a periodic-time basis, and a reactive behavior provide higher resilience while offering a better performance (overhead and energy consumption) than AODV and DSR, as shown in our simulation results.

KEYWORDS

wireless sensor networks; jamming; network resilience; routing protocols; network performance

1. INTRODUCTION

Wireless sensor networks (WSNs), sometimes covering wide geographical areas, are being deployed for a large number of applications, as they have become one of the most suitable technologies for supporting services, such as air quality monitoring, vehicular traffic monitoring, street lighting optimization, smart parking, among others [1]. In these scenarios, WSNs are exposed to time variant conditions, such as changes in the propagation of radio signals, temperature, humidity, wind, interferences and deterioration of the physical components. This random behavior affects even those WSN applications where nodes are static. Thus, the network can face from favorable to adverse conditions, and yet, it must operate uninterruptedly with the best possible performance, providing continuous information services.

In addition, other factors can impact the services delivered by the networks, such as faults and attacks. For faults, intermittent or dropped radio links can change the network connectivity. For attacks, and in particular if nodes are static, the exposure to threats can be increased, since evasion tactics are not possible [2]. Therefore, an important consideration in the design of sensor networks is the notion of resilience, where the routing protocol plays a key role in the self-configuring capabilities of the WSN to deal with adverse scenarios.

Among different types of menaces, jamming, an intentionally generated interference to degrade the services offered by a WSN, is one of the most feasible kinds of attacks. Cryptographic mechanisms cannot prevent jamming from affecting a WSN, since the interference can impact on the transmission and reception of legitimate packets. Therefore, an analysis of the effects of jamming can be useful as a way to evaluate the performance of WSNs.

Some of the parameters studied to identify the jamming nature are: resource limitations, wireless communication medium, the mobility of nodes and absence of infrastructure [3]. Another performance metric is the amount of changes in the radio links (new or dropped links) implying node connections or disconnections and, hence, a constant change on WSN topology (thus, this metric is related to network connectivity). Ultimately, high changes on the availability of routes means that the routing protocol should be able to react quickly to any changes and update routes efficiently. Additionally, when a network is disturbed, node time recovery ought to be measured [3]. Nodes' forwarding packet capability is also essential in a WSN and can be significantly affected by the characteristics of the wireless medium and noise. Nodes that carry a high amount of traffic are likely to exhaust their energy faster, and this causes network failures or traffic spikes generating packet loss. Moreover, the addition of new nodes to the network also causes changes in routing tables and topology [3]. However, it is very important to analyze the whole scenario, because sometimes, node failures or poor connectivity can have similar effects as a jamming attack [4].

1.1. Objective

The purpose of this paper is to study the effect of several types of jamming attacks (in particular, random and reactive) in a WSN. More specifically, we shall study how network performance degrades, when subject to jamming and when using any of the following routing protocols: *ad hoc* on-demand distance vector (AODV), dynamic source routing (DSR) and the multi-parent hierarchical (MPH) protocol [5,6]. Network, MAC and energy performance are captured in terms of several proposed metrics.

Throughout our research, we have found both that MPH is more resilient to jamming attacks, as well as to adverse conditions of the wireless environment in that it offers a better network performance (overhead and energy consumption) than DSR and AODV. This is both because MPH encapsulates the attack to a smaller network segment better and because of the self-configuring capabilities of MPH, derived from a combination of a periodic and proactive route update. Thus, at least for certain scenarios, there is room for using alternate protocols to AODV or DSR. Our results ultimately motivate further research in studying other network metrics that enable, in the context of MPH, a timely detection of jamming and other kinds of attacks.

We describe the organization of the rest of the paper. In Section 2, we present related work. Section 3 reviews jamming attacks. Section 4 discusses protocols, jamming and metrics analysis. Section 5 shows the description of simulator parameters and network architecture. Section 6 analyzes the results and discussion. Section 7 explains different spatial node distributions. Finally, conclusions are given in Section 8.

2. RELATED WORK

Kiwior *et al.* [7] study mobile *ad hoc* networks (MANETs); in particular, link quality due to jamming, interference, coverage range or other generic characteristics of the communication medium in an airborne environment. They analyze the connectivity and performance of known protocols using metrics, such as the packet delivery ratio (PDR), end-to-end delay and other overhead metrics.

Mingyan *et al.* [8] provide optimal detection tests that show what knowledge of a network an attacker node may acquire. The authors analyze the difficulty of detecting an attacker and the collective impact of defending a network. The authors take into account the probability of jamming and the transmission range. The work of Misra *et al.* [9], by contrast, studies a metric for measuring the degree of jamming affecting each node, providing the jammer's localization centrally. Furthermore, the authors complement the method with a decentralized setting, where each node measures the perceived jammer power in order to observe the degree of damage based on coordinates. According to the parts of the network affected by the jamming, a zoning map of the jammed area is created, as well as measures for decreasing node energy consumption are established. Likewise, the work cited in [10] studied a local perspective for detecting jamming. The metric used is throughput. The authors make a comparison between normal levels of throughput and interference levels of throughput (or abnormal). The authors establish two categories of jamming: signal strength-based and protocol behavior-based. Moreover, they try to detect the jamming affected zone (through PDR and throughput) and locate unaffected nodes to isolate them from the attack. This technique favors network performance and the study of the self-configuration feature the of routing protocol.

Xu *et al.* [4] examine jamming attacks and the corresponding defense strategies. They also analyze different methods for detecting the presence of a jammer node. The authors characterize and then analyze different types of jamming presented in a wireless medium, namely: constant, deceptive, random (which, as shall be discussed below, is a little different from our definition of random jamming) and reactive. To detect the presence of jamming in either the transmitter or receiver, the authors suggest using a number of measures, particularly signal strength, carrier sensing time, the packet send ratio (PSR) and the packet delivery ratio (PDR).

Wireless mesh networks are studied by Jiang and Xue [11]. There, the authors analyze the benefits of dynamic channel assignment, alternating routes and channels and effective routing when a network defends against a jamming attack. They also study the effect of restoration and how it affects the latency and throughput of the network. By way of comparison, in [12], Awerbuch *et al.* propose a MAC protocol capable of defending a network against a jamming attack at the physical layer. There, a model that takes into account the difference between a jamming attack or a collision is proposed, and the energy efficiency of the protocol on different attacks is shown.

By contrast, in [13], the authors analyze how powerful an attacker can become with respect to energy consumption, transmission range and coverage. They point out that the objectives of a jammer node include obstructing internode communication (especially with the collector node) and increase the energy consumption of the network node. One of the biggest challenges an attacker must face is to know the MAC protocol that the nodes run, so as to determine how one should attempt to disrupt communication.

The aforementioned works consider a number of input variables for the jamming detection mechanisms; *i.e.*, packet delivery ratio, signal strength, carrier sense metrics, *etc.* Therefore, some of the methods presented there may result in being more effective than others depending on the conditions of the attack and the scenarios. One of the motivations for this work was to conduct a comprehensive analysis on the metrics that can present changes under a jamming attack. The results of this research could be used later to create better detection methods by choosing suitable input variables. However, for a better understanding of the analysis, the underlying routing protocol should be considered as a factor that influences the behavior of the metrics. Some of the most prominent routing protocols for WSN are the AODV and DSR; since these protocols have different inner mechanisms, then the behavior of the measured variables would be different for each of them; thus, a methodology that includes a larger number of metrics than previous works and, at the same time, the influence of the routing protocol would extend some results of the state-of-the-art. Besides, the recently proposed protocol, MPH, which has design principles to deal with adverse conditions of the wireless environment, is also considered in this paper as a rival of AODV and DSR for certain scenarios.

3. JAMMING ATTACK MODELS

A jammer is defined in [14] as "...an entity who is purposefully trying to interfere with the physical transmission and reception of wireless communications". There is a number of models for jamming, and some of them will require more resources than others when they are carried out in a real scenario. Budget, effectiveness, technical complexity and energy consumption are some of the factors that are considered when defining an attack strategy.

In constant jamming, for instance, a constant emission of a radio signal is used to block a node from receiving signals from others. In addition, if that target node MAC layer is based on a carrier sense scheme, then it will not be able to transmit either, since the medium channel will always be busy. This kind of jamming produces an effective denial of service (DoS). Yet, it may require purchasing specialized hardware, such as a waveform generator or a power amplifier, since, often, a power supply or a battery with enough capacity is needed in order to provide the energy demanded by the jamming equipment. This makes jamming deployment less practical, but more importantly, it eases the identification of a suspicious device. Indeed, when a DoS becomes so evident, constant jamming can be easily detected; therefore, it might not be a reasonable long-term attack strategy.

In the process of defining how a jamming attack should be attempted, one hence may consider special hardware or even a special configuration of a wireless node. However, from the attacker's point of view, for the sake of both simplicity and inexpensiveness, the hardware that can best interfere with the channel frequencies of the target node is one that is originally intended to serve as a legitimate node of the WSN, and since, in most cases, such a device is low-cost, the attack can be very practical. In our work, we shall consider a practical scenario where the jamming is produced by a device with the same MAC and physical layer as the rest of the nodes of the WSN. Thus, the packets produced by the jamming device will have the same structure as the packets generated by legitimate nodes.

In the comparative development of this work, we have used two strategies for jamming: random and reactive. The rationale behind this selection decision is as follows. Random jamming is, on the one hand, one of the simplest types of jamming to implement and fairly difficult to detect. On the other hand, reactive jamming is both complex to implement, but rather difficult to detect. We introduce our jamming models below.

3.1. Random Jamming

We are using a random jamming model where packets are produced following a uniform distribution and ρ is the number of packets per second generated by the jammer node. As expected, ρ can be adjusted to increase or reduce the level of the jammer aggressiveness. A low value of ρ indicates a reduced interference activity, and as consequence, the DoS will not be constantly carried out. However, from a detection jamming mechanism point of view, it would be quite difficult to detect this attack, since its effects can be misinterpreted as normal interference due to the wireless environment (other wireless networks and devices sharing the same bandwidth). By way of comparison, a high value of ρ will produce a more severe DoS effect, and it could be evident that an abnormal interference activity is being performed; this scenario can be considered as a worst case for the WSN in terms of performance degradation.

We note in passing how our random jamming model compares to that of [4]. If we set ρ to a very large value, we end up with a model that Xu *et al.* call deceptive, which constantly injects regular packets into the channel (without any gap between subsequent packet transmissions). By contrast, if we set ρ to an intermediate value, then we end up with a model very similar to what they call random, the only difference being that, instead of sending a burst of two or more packets, we send only one. Our packet inter-generation time corresponds to the pause step of Xu *et al.*, where the node falls asleep.

3.2. Reactive Jamming

As in [4], reactive jamming is an attack strategy where the jammer will emit a packet with the objective of colliding with a legitimate packet. The jamming device keeps monitoring the wireless medium; once it detects a packet, it then generates a collision one immediately after. Note that the attack will be successful if at least one bit of the legitimate packet is altered, and there is no error correction scheme, only an error detection mechanism. This attack uses less energy than the random jamming strategy.

In this paper, we use rates from 50 packets/s to 80 packets/s in order to observe the network behavior under the simplest attack and that is easy to detect by the neighboring nodes, under measurements of signal strength and collision rates. On the other hand, in the other extreme, one of the most difficult attacks to detect is reactive jamming. This is because the attacker node is observing the channel in order to listen to an activity, and it causes collisions.

Table 1 presents the main characteristics and assumptions considered in this work on the implementation of two known types of jamming: random and reactive. With this table, we intend to justify the use of each of these types of jamming in order to compare the behavior of the network under each one. In addition, we will study these types of jamming by a fair comparison of two known routing protocols, AODV and DSR, and the proposed MPH, as explained earlier in [6] and evaluated under general performance metrics in [5].

Other jamming attacks are more sophisticated. Intelligent jamming is aware of the MAC or network protocols, and they can interfere with specific packets in order to maximize the jamming gain. We shall not consider this type of jamming, since we would like to focus on general jamming strategies, which can be applied despite the underlying routing and MAC mechanisms.

Table 1. Comparative study of jamming: random and reactive.

Analyzed Types of Jamming	
Random	**Reactive**
Proactive	Reactive
Simple	More difficult to detect
Higher energy consumption	Lower energy consumption
Active mode (no idle periods)	Active mode (no idle periods)
Detection with signal strength mechanisms	More complex mechanisms for detection
Transmission packet rate = 50 and 80 packets/s	When a node detects activity, it transmits

4. THE ROLE OF ROUTING PROTOCOLS TO RESILIENCE IN SELF-CONFIGURING WSNS

In WSNs, resilience capabilities can be provided by the self-configuring and adaptation mechanisms, for instance the power transmission of the nodes can be adjusted dynamically, the MAC algorithm may be aware of the wireless medium in order to optimize the performance and the routing protocol can change the topology according to the state and availability of the radio links, energy consumption, traffic balance, quality of service and other criteria. For some well-known routing protocols for WSNs, self-configuring is inherent to the design; reactive protocols, such as AODV [15] and DSR [16] update the routes when they are required; while in the MPH routing protocol [6], nodes update the routes on a periodic-time basis and when changes in the neighborhood have been detected. Despite the differences among the three aforementioned routing protocols, they use self-configuring mechanisms to build and update the topology when the conditions of the wireless environment and the available resources change. As a result, resilience to faults and attacks can be intrinsic to the above protocols. In this section, we will explain the fundamentals of the above protocols and present a brief example of their behavior under a jamming attack.

4.1. Fundamentals of AODV, DSR and MPH Routing Protocols

In AODV, a node sends an RREQ packet (route request) in broadcast mode in order to apply for a route. If a node receives this packet, it first checks if it has previously received a record of this RREQ packet. If the packet is not registered, the node retransmits it again, increasing the number of hops and creating a reverse path (the path to the node that this RREQ has reached). A node that has received an RREQ packet can only respond with an RREP packet (route reply) to confirm a route in only two cases: if the node is already the destination or if the node has an available route to the destination. This procedure continues until the source is reached. The source node may receive RREP packets from different nodes as a confirmation of various possible routes to the destination. In this case, the source node has two criteria for selecting the best route: the route that has the least number of hops or the route that has the highest sequence number. Then, nodes can have multiple routes to the same destination; the node generally chooses the shortest route.

In DSR, when a source node has a packet to send to a destination, it first looks in its cache to see if it has a route to that destination; if so, it creates a new packet header containing the path with the necessary hops to reach the destination and sends it. If there is no route to that destination in its cache, the node starts the route discovery mechanism, where the source node sends an RREQ broadcast packet, which contains the identifier of the source node and that of the destination node whose route is to be discovered; it also contains a unique identifier for the RREQ. Each intermediate node makes a copy of the received RREQ and adds its identifier. When a node receives a RREQ, it first looks in its cache to find out if it has the route to the destination node. If so, the node responds with an RREP to the source instead of forwarding the RREQ. In that RREP, the node writes all nodes that the received RREQ has passed through and the nodes of the route found in its cache. If the node does not find a route in its cache, it adds its address in the packet and forwards it via broadcast. Once the source node receives the RREP, it stores the route in its cache. This route will be included in the header of each subsequent packet that the node sends, so that all nodes that receive the packet will know to which next node they should resend. As with AODV, nodes can have multiple routes to a destination and choose the lowest hop count.

MPH is a routing protocol that creates a logical hierarchical network topology where the node hierarchy is given by its location level in the hierarchy (the lower the level, the lower the hierarchy) [6]; basically, the root hierarchy (coordinator node) will have the highest hierarchy. When a node is in a specific hierarchical level (or generation), the nodes that are connected to it and have a higher hierarchical level are parent nodes. On the other hand, nodes that are connected to any parent node and that have lower hierarchical level are child nodes. The hierarchical topology will be formed with routes that minimize the number of hops from a given node to the coordinator; this restriction will reduce the overall energy consumption of the network. The MPH routing protocol allows a child node to have one or more parents. As a result, a node can share both children and parents with another node belonging to the same hierarchical level or generation.

When a node cannot reach a specific neighbor node, then it deletes it from its neighbor table. This is a very important feature that helps to deal with the dynamic conditions of the wireless environment, such as loss of links, damaged nodes, battery exhaustion, *etc*. Besides, a proper neighbor update is especially important during a jamming attack, because this action has the effect of isolating the affected area and modifying the network topology, for instance re-routing the traffic using the rest of the network; this is true even if the attacker keeps moving. Moreover, since in the MPH protocol, the neighbor discovery packets are transmitted periodically, when the jamming attack finishes, the node can be discovered again, and it starts to behave as a normal node. This feature, intrinsic to MPH, is a proactive way to update the neighbor tables and provides to the network some resilience capabilities; this is not the case for other routing protocols, such as AODV and DSR: they may need special countermeasures against jamming attacks. In MPH, in order to assign a score to the steadiness of a neighbor, a persistence variable was defined, which is related to the number of times a neighbor node responds to the discovery packets; when a node replies to the discovery packet, it receives a maximum persistence value, and for each time the neighbor does not respond, then it will be decreased by one. If the persistence variable reaches zero, then it is concluded that the node is not responding, and it is deleted from the neighbor table.

The MPH protocol mainly uses three packets: the *ND* (network discovery), the *NDR* (neighbor discovery response) and*NDRACK* (neighbor discovery response ACK) packets. At initial state, if the node is the coordinator, then we set its hierarchy to the maximum value; otherwise, we simply set it to zero. Then, each node sends periodically *ND* packets to initiate the discovering and updating of its neighbors.

When a node receives an *ND* packet, then it answers with an *NDR*packet, which includes its hierarchy information. The node that receives an *NDR* packet will reply with an *NDRACK*acknowledgment packet that also includes its hierarchy level. When an *NDR* or *NDRACK* packet arrives, the updating of the hierarchy for a node except the coordinator is summarized as shown below:

```
1. Receive packet from neighbor.
2. Update neighbor table.
3. {If} neighbor_hierarchy > this_node_hierarchy + 1:
  a. Set the neighbor as the parent (next node in the neighbor table);
  b. It assigns itself a hierarchy level smaller than its parent:
  this_node_hierarchy = neighbor_hierarchy − 1.
  c. Send a ND packet to advertise its hierarchy in the neighborhood.
4. If neighbor_hierarchy == this_node_hierarchy + 1,{then} set the
neighbor as another parent.
```

When a node has data to transmit, then it creates an MPH packet and sends it to a parent node; if the parent node that receives the packet is the coordinator, then the packet has reached its destination; otherwise, the packet will be forwarded to any of the parent nodes.

Figure 1 shows the most important features of the AODV, DSR and MPH protocols.

4.2. Example of Self-Configuring Mechanisms in WSNs

Figure 2 shows an example of the link configuration for a given WSN running the AODV, DSR and MPH protocols, this example consists of three phases: in the initial phase, the nodes are switching on; in the second phase, the network has been formed according to the underlying routing protocol; and in the third phase, we can see the resultant topology under a reactive jamming attack. In the second phase, each routing protocol establishes the links according to their rules. For AODV and DSR, links are established within the coverage radius of the nodes. Regarding MPH, parent-child links are set with a hierarchical topology with respect to a given coordinator node (node zero). The third stage presents how the topology for AODV, DSR and MPH is reconfigured after an attack caused by a jammer node represented by the shaded node. Let us assume that this node produces the reactive jamming (which is the worst case scenario for the WSN presented in this paper) and causes the links next to it to become unusable. The logical links are dropped due to the partial overlapping of the coverages of the jammer node with Nodes 2, 4 and 5 having in common the lost links. This situation is caused because the attacker node emits a packet when it listens to any packet transmission, causing collisions and corrupting the legitimate packet. The nodes under the MPH protocol update their neighbor tables as soon as they realize that there are no packets being received from some neighbors. In AODV and DSR, routing tables are also updated through error messages when the links are already dropped. The nodes in the WSN with AODV and DSR increase the average number of packet retransmissions and the number of MAC algorithm retries, because the jammer node constantly maintains the reactive attack. After the topology update, these links are lost, and error packets of these routes are received by the nodes. For MPH, nodes quickly realize the futility of these affected links due to persistence and reactivity mechanisms [6]. In addition, the periodicity of the neighbor tables' update produces these links that are discarded without sending extra packet notification. Finally, this reflects lower power consumption and better performance for a network under the MPH protocol.

5. DESCRIPTION OF SIMULATOR PARAMETERS AND NETWORK ARCHITECTURE

These experiments are based on a grid of 49 nodes, where 48 of them send data with a rate of 1%–4% packets/s to a coordinator node (in the upper-left corner) functioning as a sink node. The grid is shown in Figure 3. The network consists of a MAC layer under the CSMA/CA (Carrier Sense Multiple Access with Collision Avoidance) protocol, with a maximum of three retransmissions per packet and maximum of five CSMA retries, while the nodes are waiting for the channel to be idle before transmission. The packet rate of the jammer node varies according to the type of jamming attack; random (ρ = 50 packets/s), random (ρ = 80 packets/s) or reactive.

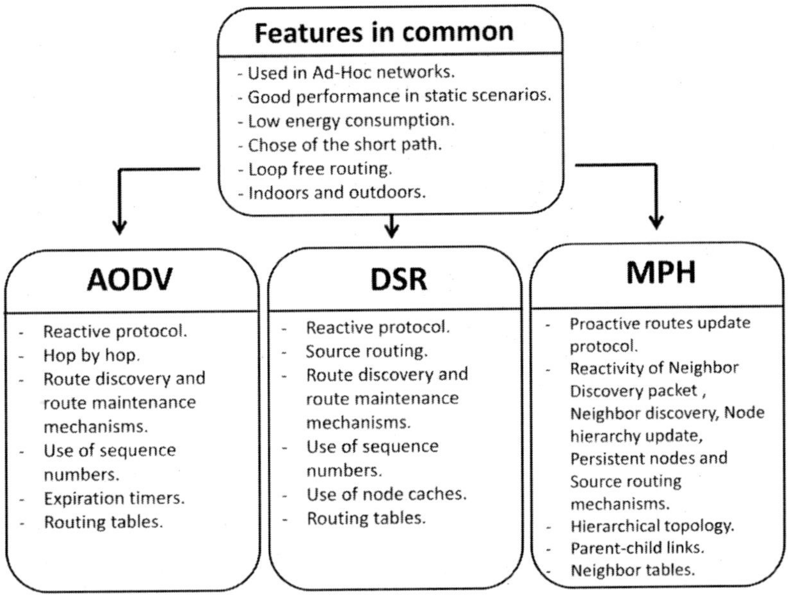

Figure 1. Important features of the *ad hoc* on-demand distance vector (AODV), dynamic source routing (DSR) and multi-parent hierarchical (MPH) protocols.

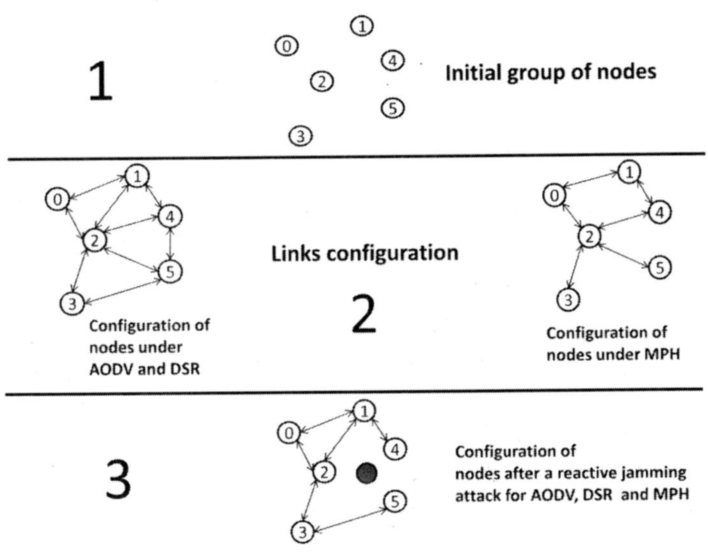

Figure 2. Self-configuring WSNs: AODV, DSR and MPH under a reactive jamming attack.

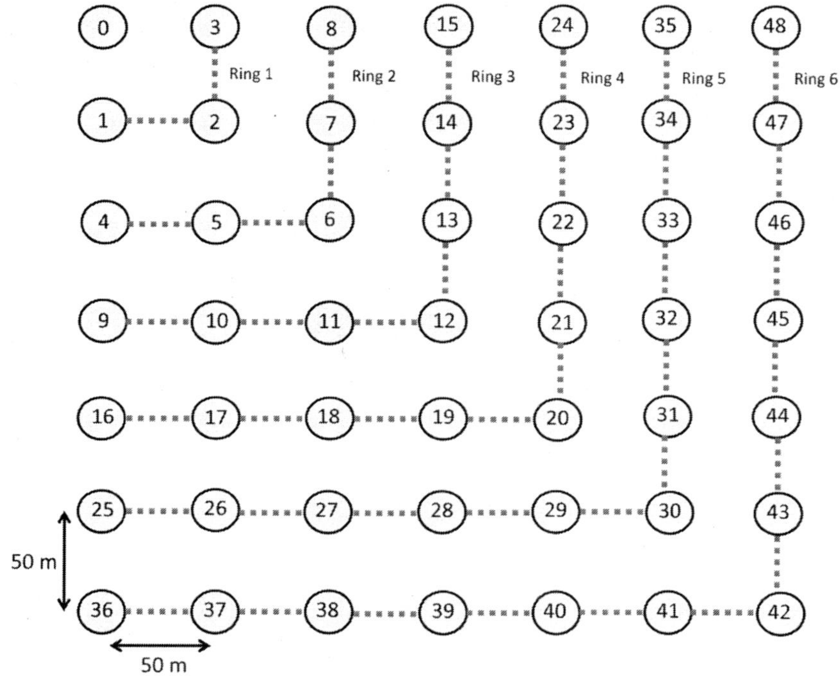

Figure 3. Network topology.

Let us consider using a transmission power of 0 dBm [17] for active transmission mode according to the CC2530 chip [18] and a receiving power of −85 dBm according to the IEEE standard 802.15.4 [19]. With these power values, we obtain an average coverage radio of 50 m for sensors following Equation (1) for outdoors [19]:

$$d = 8 * 10^{\frac{P_t - P_r - 58.5}{33}} \tag{1}$$

where d is the coverage radio, P_t is the transmission power in dB and P_r is the receiving power in dB.

Table 2 shows the physical, MAC and network layer parameters that were used in the simulator.

Table 2. Simulation parameters. CSMA, carrier sense multiple access.

Parameter	Value
Physical layer parameters	
Sensitivity thresholds	−78 to −94 dBm
Transmission power	0 dBm
MAC layer parameters	
Maximum retransmission number	3
Maximum retry number	5
Maximum number of tries to reach a node from the coordinator	9
Packet error rate	1%–4%
Average frame length	22 bytes
Maximum number of backoffs	4
MAC protocol	IEEE 802.15.4
MAC layer	CSMA/CA
Network layer parameters	
Number of nodes	49
Discovery neighbor time	10 s
Maximum data rate	250 kbps
Scenario	Static nodes

5.1. Performance Metrics and Sampling Periods

The retransmissions metric is defined as the number of times to retransmit a packet because the source node does not receive an acknowledgment. We establish a maximum of three retransmissions for each packet; once this value has reached, the packet is dropped. This metric indicates the link state of the network and can provide insight connections and the connections of nodes. The lower this metric value, the better.

The CSMA retries metric is the number of times a node runs the inner loop of the CSMA/CA algorithm. When a node is about to transmit a packet, it first listens to the channel and notes if the channel is available for transmission (carrier sense). If the channel is busy, then the node waits a random delay for another carrier sense. This cycle is what we call CSMA retry. This parameter has a maximum default value of five times in the IEEE 802.15.4 standard [19]; once this number has been reached, the packet is discarded. This metric gives an indication of the wireless medium state that is influenced by the amount of packet collisions that force repeated CSMA backoffs. The lower this metric value, the better.

The tries metric is defined as the number of times the coordinator node needs to communicate with every single node in the network. This metric also is known as route availability. The coordinator is a robust node and has powerful processing capability to be aware of the entire network topology. We set a maximum of nine tries for each node. This metric relates connectivity to the link state and topology changes. The lower this metric value, the better.

For the metrics retransmissions and CSMA retries, we set a sampling period of 100 s, where for the first 30 s, the network is in normal mode. Then, in the next 30 s, a node jammer is introduced in any of the positions evaluated (near, medium or far from the node coordinator, which is always the destination node). Finally, for the next 40 s, the jammer node is removed, and the network returns to a normal state.

For the tries metric for connecting from the coordinator to each node, we established 100 s of normal network mode and 100 s in the presence of a jammer node. Then, we run this "ping" process for every single node of the network by the coordinator, so this constitutes approximately 20 repetitions of the process in a period of 100 s for significance in results. We insist that we have set the maximum tries to nine per node.

The achievable nodes by the coordinator metric is defined as the number of nodes that can answer the queries from the coordinator. Ideally, the coordinator must know the whole network topology and, therefore, should be able to reach all nodes when required. The higher this metric, the better for the network. It is important to note that this metric is related to the tries metric with each of the nodes. We have set a maximum of nine tries to reach a node.

Energy is a metric that quantifies the consumption of resources on each node. When a network has low performance, the energy consumption will be higher, because the number of collisions increases and, thus, retransmissions and CSMA retries. Furthermore, when the routing protocol is not able to recover quickly from an attack, the link state and energy consumption will be increased. For assessing this metric, we established 100 s of normal network mode and 100 s in the presence of a jammer node.

6. RESULTS, ANALYSIS AND DISCUSSION

Results of the studied variables are displayed by statistical graphs and 2D spatial distribution figures. Statistics help us to note how the most relevant information on nodes is organized. The spatial distribution gives us an idea of which nodes or node zones are affected and what the impact is in the whole network. Tables that accompany the following results summarize the information contained in the figures presented in this section and help to compare the results for each of the protocols.

6.1. Statistical Analysis of Performance Metrics

The methodology used to present the following results is the box and whisker method, which shows the distribution of data for each of the evaluated metrics and allows us to have a representation that describes the dispersion and symmetry characteristics. It represents three quartiles (Q_1, Q_2, Q_3) and the minimum and maximum data values on an aligned rectangle, horizontally or vertically. Q_1, the first quartile, is the value greater than 25% of the values of the distribution. Q_2, the second quartile, is the median of the distribution; it is the value of the variable that takes center place in a set of ordered data. Q_3, the third quartile, is the value that exceeds 75% of the values of the distribution. The interquartile range is Q_3–Q_1, and it represents 50% of the nodes. In order to develop this method, we followed the standard approach described in [20].

For the purpose of the results analysis, it is important to note that the graphs of the box and whisker method has been built with a whisker coefficient of 1.5. The extremes in each box represented by a line show the maximum and minimum values of observed data. The ends represented by X are the 1% at the bottom and the 99% at the top of the box. The filled squares shown in the boxes represent the mean of the data.

6.1.1. Retransmissions with Box and Whiskers Representation

Figure 4 shows the average retransmissions for the AODV, DSR and MPH protocols for each of the attacks studied in this work; random (ρ =

50, 80 packets/s) and reactive. The figure presents the position when the jammer node is in the middle of the topology.

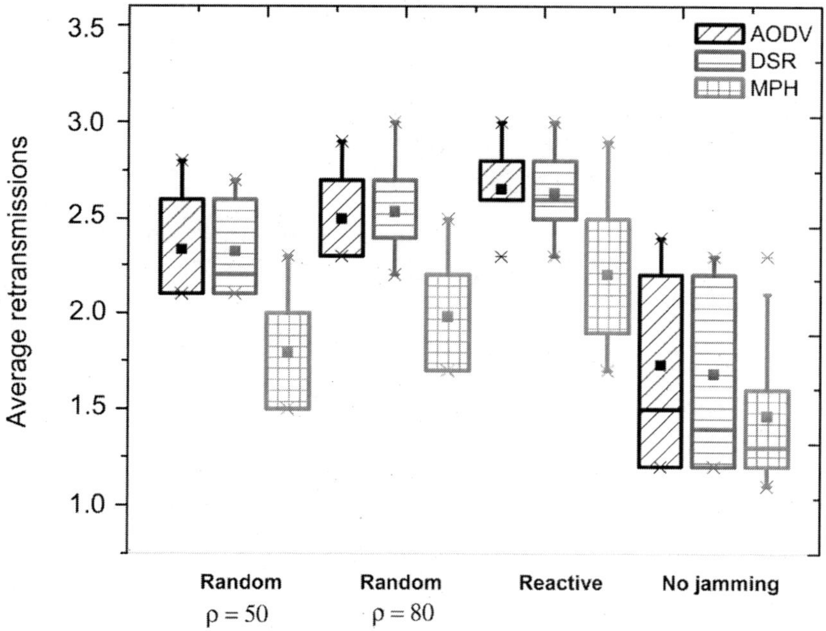

Figure 4. Average retransmissions when the jammer node is in the middle of the topology.

For the retransmissions metric with no jamming activity, the MPH protocol has a smaller box, which means that the values of average retransmissions are below the values that present the AODV and DSR protocols. Then, the MPH protocol has 15% less average retransmissions than AODV and 13% less than DSR. We observe that when there is no presence of a jammer node, the bottom of the box is smaller than the top. This means that average retransmissions values between 50% and 75% of the data population are more dispersed than data between 25% and 50%.

Figure 4 describes the three different attacks for each of the studied protocols with the jammer node in the middle of the network topology. Note that in the three jamming types, the boxes are short in size, so that we can infer that when the jammer node is in the middle of the topology, it affects more nodes, because it can cause an attack that spreads in the concentric circles form. However, boxes for MPH indicate that the WSN can be resistant to this wave of attack, because their values are significantly lower than AODV and DSR. With respect to the random jamming with $\rho = 50$ packets/s, MPH has average values of 23% lower than AODV and DSR. Regarding the random jamming with $\rho = 80$ packets/s, MPH shows an improvement of 20.5% over AODV and 21.7% over DSR. Under reactive jamming, MPH reacts 16.7% better than the other two studied protocols, and we note again that under the most aggressive network attack, the lowest number of nodes affected by the jamming is when the MPH is used.

In what follows, we explain the results for both cases, when the jammer node is near and far from the coordinator. For the near case, we obtained that over the network nodes, there are significant differences in the values of this metric. Moreover, the distance or proximity of the nodes with respect to the coordinator influences collisions, packet loss and, therefore, retransmissions. This is because the nodes that are closer to the coordinator node forward more traffic than the distant nodes. Furthermore, the presence of the jammer node close to these nodes can cause bottlenecks and more collisions. However, the MPH protocol has an average of retransmissions of 28% less than AODV and DSR, for random ($\rho = 50, 80$ packets/s) and reactive jamming. One possible explanation for this fact is that the AODV and DSR protocols affect more nodes under the most aggressive type of jamming (reactive), while in MPH, fewer nodes note the attack.

For the far case, we obtained that for a random jamming with $\rho = 50$ packets/s, we notice that MPH has 30% lower average retransmissions than AODV and DSR. With respect to a random jamming with $\rho = 80$ packets/s, MPH has 15% lower average retransmissions than AODV and DSR. With respect to reactive jamming, the average retransmissions values are still lower for MPH; for instance, the average for MPH is less than 13% against AODV and 15.4% with respect to DSR.

6.1.2. CSMA Retries with Box and Whiskers Representation

The CSMA retries metric is presented in Figure 5. When there is no jamming, we can notice that the MPH protocol has an average of 2.3 of CSMA retries, while AODV has a mean of 2.86 and DSR 2.76. In addition, the MPH box shows that more than 50% of the nodes have average values between 1.38 and 2.7 (interquartile value), while AODV keeps 50% of the nodes between 2.2 and 3.3 average CSMA retries and DSR between two and 3.21. With these values, we can realize that MPH has fewer average CSMA retries and a better performance compared to the other two protocols.

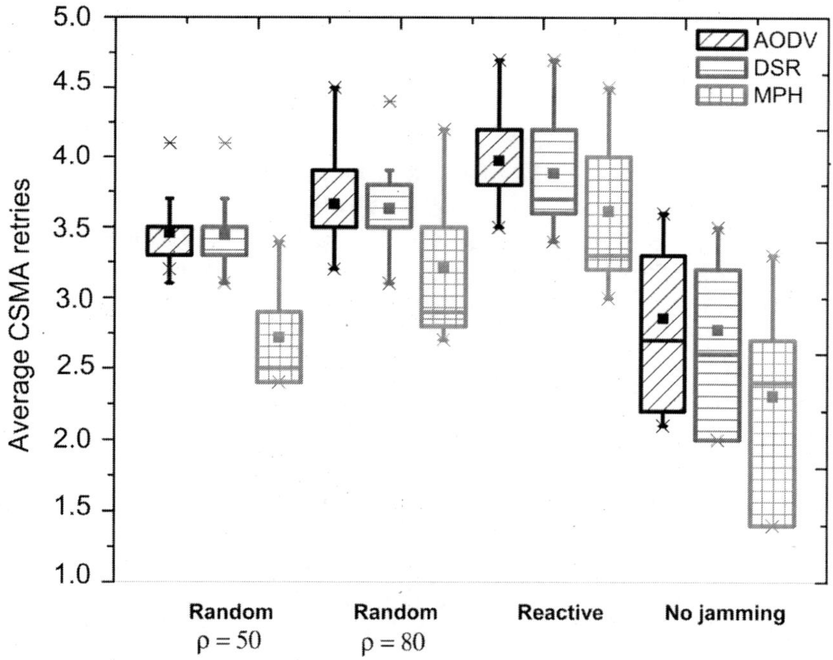

Figure 5. Average CSMA retries when the jammer node is in the middle of the topology.

Figure 5 exposes the types of jamming when the jammer node is in the middle of the topology. This jammer node location may affect many nodes, but its impact in the total delivery packet ratio should not be as strong as it is when the attacker node is closest to the destination node. With respect to the random jamming (ρ = 50 packets/s), AODV and DSR have an interquartile data distribution between 3.3 and 3.5 of the CSMA retry average values. This means that the nodes next to the coordinator deal with a more busy wireless medium due to the higher traffic load compared to distant nodes; therefore, the values in the CSMA retry parameter is greater than in other network nodes. Besides, if we consider that the central part of the topology is being affected by an attacker node, then almost all nodes in the network will have similar values regarding the evaluated variable. This also reveals that AODV and DSR have more affected nodes than MPH. This is because, for MPH, the box is slightly elongated, with an interquartile between 2.4 and 2.9. Furthermore, these values are significantly lower than in the case of AODV and DSR. For cases of random (ρ = 80 packets/s) and reactive jamming, the behavior is similar to the described jamming above, where MPH has lower values for the evaluated variable, and elongated boxes show that the affected nodes by the jammer are fewer than in the AODV and DSR cases.

Regarding the case when the jammer node is close to the coordinator and considering conditions of random jamming withρ = 50 packets/s, MPH has 18.4% lower average CSMA retries compared to DSR and AODV. For a random jamming with ρ = 80 packets/s, MPH has 7.8% lower average CSMA retries. Finally, for reactive jamming, MPH is better by 8.8% than DSR and by 16.4% than AODV.

For the case when the jammer node is far from the coordinator node, we note that for random jamming with ρ = 50 packets/s, MPH has 40% less average CSMA retries with respect to the results for AODV and DSR. For a random jamming with ρ = 80 packets/s, MPH has 50% less average CSMA retries with respect to AODV and DSR. Finally, in the reactive jamming, MPH has 61% less average CSMA retries than the values obtained for AODV and DSR. This can be explained considering that the MPH protocol is not strongly affected when the attacker node is away from the destination. Although the average values of CSMA retries increase, MPH is able to encapsulate the attack, affecting only a few nodes. AODV and DSR tend to affect more nodes, and these values represent an increment in the average of CSMA retries in the nodes nearby the coordinator.

6.1.3. Tries with Box and Whiskers Representation

In Figure 6, we display the average number of tries performed by the coordinator node to communicate with each node in the network. This variable indicates how much knowledge the coordinator node has about the network topology and the quality of the routes. When there is no jamming, the protocols have boxes of the same size for the three cases, which means that 50% of the data is distributed in approximately the same average values. However, the average value of tries for MPH is

1.33, whereas for DSR, it is 1.58 and for AODV 1.65. This means that MPH has 15.8% better performance connectivity of nodes with respect to DSR and 19.4% compared with AODV. Then, 50% of the nodes has between one and two tries, 25% of the nodes three tries and 25% of the nodes four tries for DSR. Besides, 50% of the nodes has between one and two tries, 25% of the nodes three tries, 24% of nodes four tries and 1% of the nodes five tries for AODV. While for MPH, 50% of the nodes has between one and two tries and 50% of nodes three tries (three being the maximum value reached in MPH under normal network behavior).

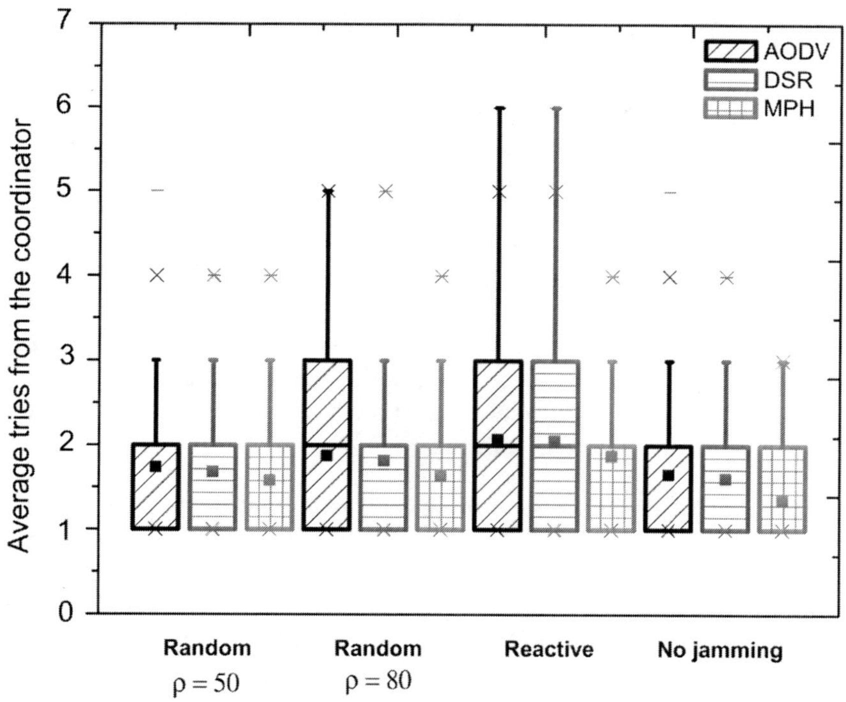

Figure 6. Average tries when the jammer node is in the middle of the topology.

Figure 6 denotes metric tries when the jammer node is in the middle of the topology. We notice that in the random (ρ = 50 packets/s) jamming, the boxes of the three protocols remain the same as when there is no jamming (this is equivalent to 50% of nodes). However, for AODV, some nodes manifest variations on the evaluated parameter, which is 24% of nodes having four tries and 1% of nodes having five tries. For DSR and MPH, 50% of the remaining nodes presents four tries. However, for MPH, the average value of tries is 1.57, while for DSR, it is 1.7 tries. For random (ρ = 80 packets/s) jamming, AODV changes, and now, 50% of the nodes is between one and three tries and 50% of nodes four or five tries. DSR maintains 50% of nodes between one and two tries, another 25% of nodes three tries and the other 25% between four and five tries. MPH's percentages of nodes are maintained the same as in the random (ρ = 50 packets/s) jamming. Concerning the reactive jamming, AODV and DSR have 50% of nodes from one to three tries, and the other 50% of nodes has four, five and six tries. MPH maintains its previous features only with a small variation in the mean value of tries. With these numbers, we remark that the MPH protocol maintains its good performance and is not prone to collapse due to faulty links.

With respect to the case when the jammer node is close to the coordinator, for the random (ρ = 50 packets/s) and random (ρ = 80 packets/s) jammings, MPH has 12% low average tries with respect to AODV and 10% against DSR. One notable change can be observed in the reactive jamming, where AODV and DSR vary from one to three tries for 50% of the nodes. AODV varies between four and five tries for the other 50% of nodes, and for DSR, the remaining 50% of nodes has four tries. MPH maintains 50% of nodes varying between one and two tries, 24% of nodes three tries and only 1% of nodes four tries. This result shows the strength of the MPH against an attack and its better performance regarding encapsulation of failures.

For the case when the jammer node is far from the coordinator node, for the random (ρ = 50 packets/s) jamming, the three protocols have 50% of nodes with values between one and two tries and 25% of nodes three tries. For AODV, the remaining 25% of nodes has five tries, and the tries' average value is 1.8. In DSR case, the remaining 25% of nodes has five tries, and the tries mean value is 1.72. In contrast, for MPH, the remaining 25% of nodes has four tries, and the tries mean value is 1.6 (the lowest value compared with the other protocols). In the random (ρ = 80 packets/s) jamming, AODV has 50% of nodes between one and three tries, while DSR and MPH remain with the same values in the random (ρ = 50 packets/s) jamming case. For AODV, the remaining 25% of nodes has five tries, and the tries' average value increases to 1.95. In the DSR case, the remaining 25% has five tries, and the tries' average value increases to 1.87. In contrast, for MPH, another 24% of nodes has four tries and the remaining 1% five tries, and the tries' average value is 1.76. Finally, in the reactive jamming case, AODV and DSR have 50% of nodes between one and three tries, while MPH values remain unchanged (between one and two tries). The AODV and DSR cases have a very similar behavior, where the remaining 25% of nodes has five tries, and the tries' average value increases to 1.97. On the other hand, for MPH, the remaining 25% of nodes has four tries, and the tries' average value is 1.82. This MPH behavior describes that although the random (ρ = 80 packets/s) and reactive jammings affect MPH nodes similarly, MPH exposes better performance and link connectivity than AODV and DSR protocols.

6.2. Analysis Spatial Distribution of Performance Metrics

In this part, we show the 2D spatial representation of the studied performance metrics and exhibit the number of reached nodes by the coordinator in order to expose the self-configuration capacity of each protocol.

6.2.1. Retransmissions with 2D Representation

Figures 7–9 present the worst case of jamming, the reactive case, for the three different positions of the jammer node. Each figure shows a grid of nodes that represents the used topology, a matrix of 7 × 7 nodes. Through a grayscale, we exemplify the maximum (close to black color) and minimum (close to white color) values. Grids of the "no jamming" column show the normal behavior of the network under no attack for each of the studied variables. The other three columns in each graph show the reactive case when the jammer node is close to (near column), when it is in the middle of the topology (middle column) and when it is far (far column) from the coordinator node. These grids have the change in the reactive jamming for each of the nodes, *i.e.*, the difference between the reactive jamming for each case and the value when there is no jamming for each of the nodes.

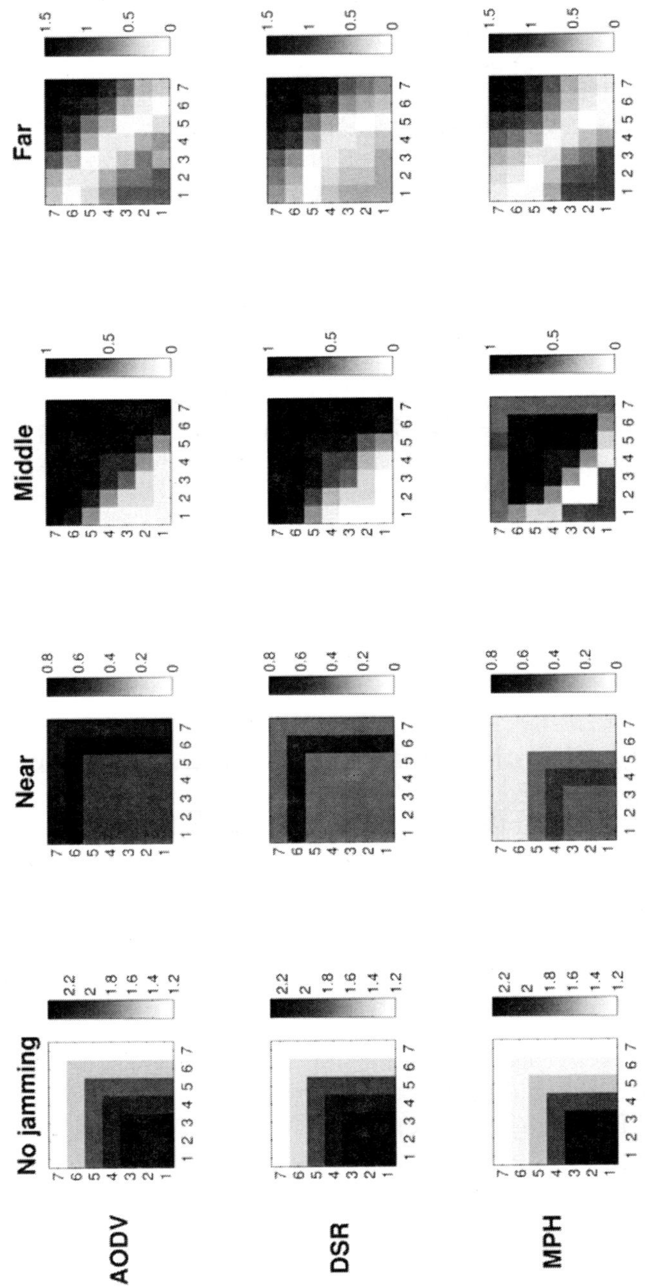

Figure 7. 2D representation of the average retransmissions in the no jamming case and the difference between the values in the reactive and no jamming cases for each of the positions of the jammer node.

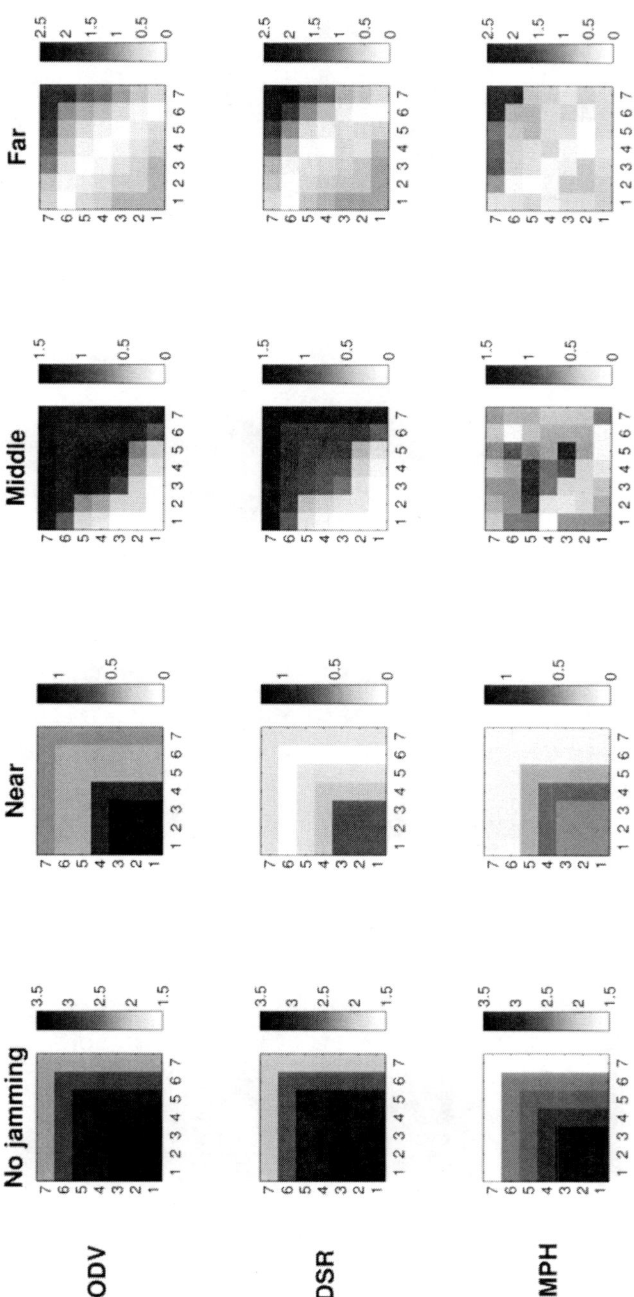

Figure 8. 2D representation of the average CSMA retries in the no jamming case and the difference between the values in the reactive and no jamming cases for each of the positions of the jammer node.

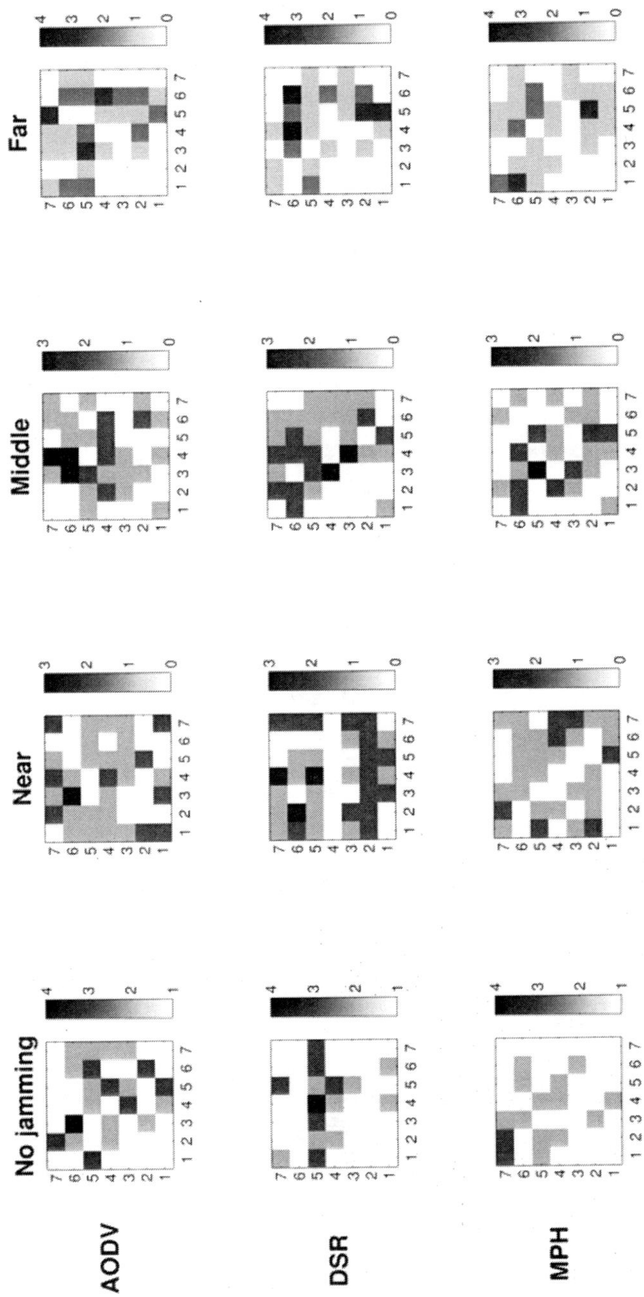

Figure 9. 2D representation of the average tries in the no jamming case and the difference between the values in the reactive and no jamming cases for each of the positions of the jammer node.

In Figure 7, we note that in the no jamming column, we have the average number of retransmissions for the AODV, DSR and MPH protocols. In these grids, we notice that there is a greater amount of retransmissions in nodes near the coordinator, because these nodes generate and forward traffic from other nodes. For this reason, collisions are more likely to occur and, hence, packet retransmissions. However, the MPH protocol has lower average retransmissions, and the nodes that are affected by collisions are less than AODV and DSR cases. Concerning the near column, we observe the difference in values between the reactive and no jamming case for each node. We note that for AODV, the biggest change is in the farthest nodes to the coordinator, because the jammer node is close to the coordinator, and this prevents the final communication, making the number of affected nodes greater in the network. Then, this attack position alters nodes that are far from the destination for AODV. Therefore, nodes close to the coordinator node have a change between 0.4 and 0.6, which shows that they are also significantly affected by the attack. For DSR, the attack is a little less aggressive, but also affects one of the farthest nodes crowns with respect to the coordinator with a change of about 0.6. The jamming has the lowest impact in terms of the number of nodes for the MPH; the change for nodes that are closer to the coordinator is approximately 0.4. Distant nodes have a change between 0.1 and 0.2. Regarding the middle column, AODV and DSR modified approximately one half of the nodes in the topology, while MPH encapsulates the attack slightly more with less affected nodes. With regard to the far column, AODV and DSR for the jammer node have more impact with a change of 1.5 in the near node jammer nodes. The MPH protocol has a change of one in about three nodes directly connected to the jammer node, and then, the change tapers to one, *i.e.*, MPH does not allow the attack to spread to multiple nodes as AODV and DSR.

6.2.2. CSMA Retries with 2D Representation

Figure 8 shows the average CSMA retries when there is no jamming and reactive jamming cases when the jammer node is near, in the middle and far with regard to the coordinator. Note that the no jamming column for AODV has a maximum of 3.5 average CSMA retries for the closest nodes to the coordinator, while DSR has an intermediate value between three and 3.5, and MPH a value of three. We remark that as nodes move away from the coordinator, the CSMA retries are fewer, because these nodes forward less traffic, and therefore, the channel is less congested for communication. MPH reaches a value of 1.5 CSMA retries to the farthest nodes from the coordinator, while DSR has a value of two and AODV a value between two and 2.5. In the near column, we clearly observe that AODV has the highest difference (about one) for the nearest nodes to the coordinator. DSR continues with a difference value between 0.5 and one for these nodes and, finally, MPH with a value of 0.5. We note that MPH has the smallest change for the farthest nodes from the coordinator, showing that collisions caused by the jammer node do not affect a more extensive area in a large scale, as in the AODV and DSR cases. With respect to the middle column, the biggest change is observed for the AODV protocol, with values between one and 1.5 for about half of the nodes (the farthest part from the coordinator). DSR has a slight change between 0.5 and one for about the same half of nodes in the topology. While MPH has no abrupt changes in network nodes, it has only a few nodes with values between 0.5 and one around the central part of the topology (just around the jammer node position). The far column presents a similar behavior for the three protocols; however, MPH is the one with the least change (values between 1.5 and two) in the closest nodes with respect to the jammer node, and it is important to remark that although AODV presents the same values, MPH covers fewer nodes than AODV with these values.

6.2.3. Tries with 2D Representation

Figure 9 exhibits the tries metric when there is no jamming in the network and reactive cases where the jammer node position is near, in the middle and far with respect to the coordinator. In the no jamming column, we observe that for all three protocols, nodes that are close to the coordinator needed fewer tries to be reached (one try). AODV and DSR reach up to four tries, while the maximum tries for MPH is three. For reactive jamming, cases in the other columns are not well defined in any of the protocols with respect to a specific pattern; however, MPH presents shaded nodes with lighter colors than the AODV and DSR cases, representing that the differences in the tries metric under reactive attack are softer than in the AODV and DSR protocols.

Table 3 helps us to visualize and quantify more easily the results that we observed in Figures 7, 8 & 9. In this table, we can perceive the three studied metrics (retransmissions, CSMA retries and tries). Numerical values represent the total sum of the difference values in each of the nodes for each jammer node position. These values show what protocol presents the biggest changes for the worst case (reactive jamming) better. We note that in most cases, the MPH has the smallest sum, meaning that the protocol is less prone to change when a reactive jamming exists in the network and that nodes are able to diminish the attack or focus it in a small region of the network.

Table 3. Sum of the values of differences for Figures 7–9.

Retransmissions			
	Near	**Middle**	**Far**
AODV	30.4	30.7	32.6
DSR	22.9	36	31.5
MPH	13.1	31.4	28.6
CSMA Retries			
	Near	**Middle**	**Far**
AODV	32.5	42.7	31.5
DSR	12.8	39	36.9
MPH	14.7	25	31.1
Tries			
	Near	**Middle**	**Far**
AODV	42	39	41
DSR	53	43	38
MPH	37	34	34

6.2.4. Retransmissions, CSMA Retries and Tries with 2D Representation for 225 Nodes

In order to check the validity of our results and the behavior of the protocols, we simulated a network with more nodes.Figure 10 shows the results for a grid of 225 nodes. Retransmissions, CSMA retries and tries are simulated: two columns for each of these studied metrics. The left column in each case represents the case when there is no jamming, and the right column shows the difference between the no jamming case and the reactive jamming case when the position of the jammer node is at the center of the topology. These results are presented for the three protocols: AODV, DSR and MPH. The results expose that the performance of the three protocols tends to be similar when the number of nodes increases, so we can confirm that our grid of 49 nodes really explains what happens in the network. The AODV protocol presents the greatest average number of retransmissions, CSMA retries and tries. The MPH protocol offers the best resistance to the attack and causes fewer nodes to be affected. These results can be quantified in Table 4 to specifically observe the exact total numbers.

6.2.5. Number of Reached Nodes by the Coordinator

Tables 5–7 show the number of reachable nodes by the coordinator for each of the positions of the jammer node. The column designated by the time, in seconds, presents periods of 10 s where there is an average count of how many nodes were reachable by the coordinator. The shaded rows in the tables show the times in which the attack is presented as random (ρ = 50 packets/s), random (ρ = 80 packets/s) or reactive case. As we can see, the position when the jammer node is in the middle of the topology is the one that most affects the total number of nodes that responded to the coordinator. This phenomenon may be caused due to the fact that when the jammer node being in the middle, interfering communication, it has complete circles of nodes around itself and may affect more of them. Whereas the jammer node is in one of the corners of the topology (near and far positions), it has incomplete circles of nodes; however, the far position affects fewer nodes than the nearest position. This may occur because an attack is stronger in a region with a high traffic concentration, such as the area near the coordinator node, because these nodes generate traffic and also forward traffic from other nodes to the coordinator node (destination). This causes a larger amount of collisions, and this can be increased if there is an attack present.

6.2.6. Energy

In Figure 11, the total energy at each node is shown in the case of reactive jamming. This is analyzed for the three positions of the jammer node: near, in the middle of the topology and far from the coordinator. The model of the energy is shown in Table 8, used for the Texas Instruments CC2530 chip. We can see that when there is jamming, the concentration of power is in the next coordinator nodes, because these nodes generate traffic and also carry traffic from other network nodes.

However, the MPH protocol saves about 46% of the energy of these nodes closest to the coordinator. Furthermore, when we analyzed the total energy for the case when there was no jamming, the MPH protocol represents a 42% energy savings compared to AODV and 37% compared to DSR. Using Table 9, we have the quantization of the total energy for each of the protocols in each of the positions of the jammer node. We can see that when the jammer node is in the center of the attack, the topology appears to be more aggressive with respect to energy consumption, that is at the cost of a greater number of nodes. We note that the MPH protocol consumes less energy in all cases regarding AODV and DSR. For the close position, MPH has a savings of 44% compared to 41% AODV and DSR. For the center position, MPH has a savings of 41% compared to 38% AODV and DSR. For the far position, MPH has a savings of 43% compared to 42% AODV and DSR.

6.2.7. Impacts of Different Attack Strategies on the Routing Performance

Table 10 reflects the impact of different attack strategies on the routing performance. The metrics studied throughout this work show the performance of each routing protocol (AODV, DSR and MPH). This table summarizes the positive (↑) or negative (↓) impact of better or worse performance on MPH against AODV and DSR. Here, we observe the types of studied jamming and their influence on each of performance metrics. This gives us an idea of what type of jamming is more aggressive in a network when the jammer is located around the center of the topology and the coordinator node (final destination) is at the corner.

7. DIFFERENT SPATIAL NODE DISTRIBUTIONS

In this section, the main parameters of MAC and network layers for simulations of three different arrangements of nodes are described. Table 11 exhibits the parameters to simulate the three mentioned scenarios.

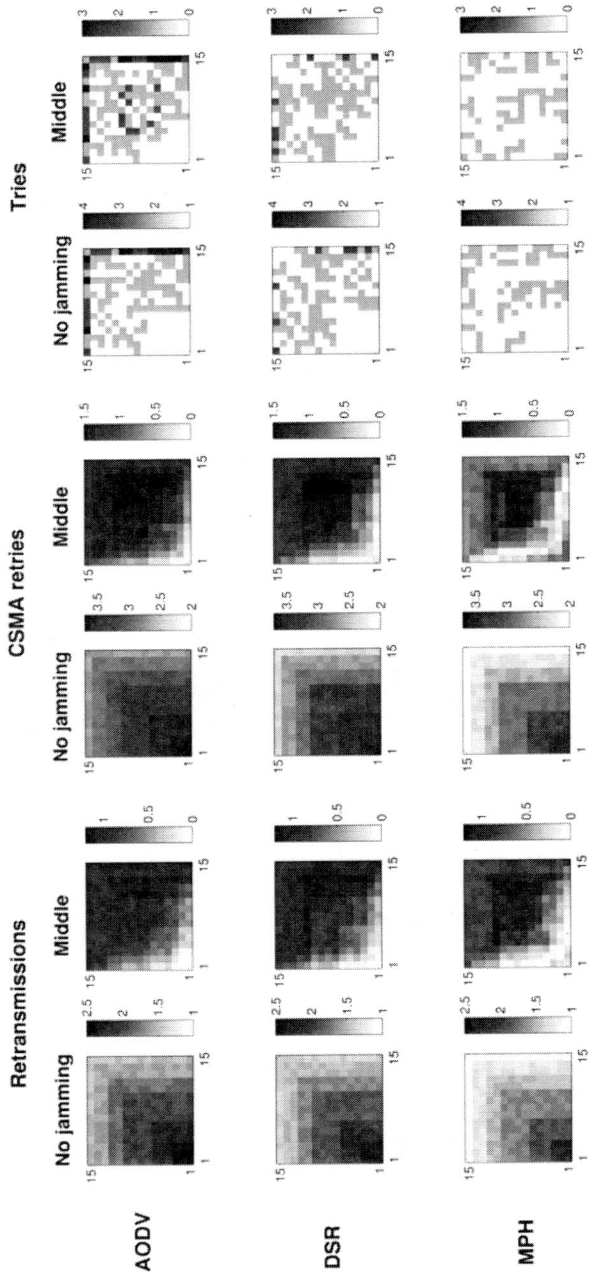

Figure 10. 2D representation of the average retransmissions, CSMA retries and tries in the no jamming case and the difference between the values in the reactive and no jamming cases when the jammer node is in the middle of the topology.

Table 4. Sum of the values of differences for Figure 10.

Position of the Jammer Node = Middle			
	Retransmissions	CSMA Retries	Tries
AODV	186.2	265.2	159
DSR	175.9	227.4	102
MPH	152.1	173.1	83

Table 5. Number of reached nodes by the coordinator when the jammer node is near.

Time	Number of Reached Nodes								
	Random ($\rho = 50$ packets/s)			Random ($\rho = 80$ packets/s)			Reactive		
	AODV	DSR	MPH	AODV	DSR	MPH	AODV	DSR	MPH
10	43	44	44	44	43	44	44	45	46
20	47	48	48	48	47	47	47	48	48
30	47	48	48	48	47	48	48	48	48
40	46	46	46	47	45	46	46	46	47
50	41	43	46	41	40	44	39	40	44
60	41	43	46	41	40	44	39	40	44
70	41	43	45	40	41	45	42	41	45
80	46	47	48	47	47	46	44	45	46
90	47	47	48	48	47	48	46	46	48
100	48	48	48	48	48	48	47	47	48

Table 6. Number of reached nodes by the coordinator when the jammer node is in the middle of the topology.

Time	Number of Reached Nodes								
	Random ($\rho = 50$ packets/s)			Random ($\rho = 80$ packets/s)			Reactive		
	AODV	DSR	MPH	AODV	DSR	MPH	AODV	DSR	MPH
10	44	44	44	43	43	45	44	45	46
20	47	47	48	48	47	47	47	48	48
30	47	48	48	48	47	48	48	48	48
40	44	44	46	43	43	46	41	41	45
50	40	40	45	39	38	44	38	38	44
60	40	40	45	39	38	44	38	38	44
70	41	42	45	40	40	44	40	41	45
80	45	47	48	47	47	46	44	45	46
90	47	47	48	47	47	48	46	46	48
100	47	48	48	48	48	48	47	47	48

Table 7. Number of reached nodes by the coordinator when the jammer node is far.

Time	Number of Reached Nodes								
	Random ($\rho = 50$ packets/s)			Random ($\rho = 80$ packets/s)			Reactive		
	AODV	DSR	MPH	AODV	DSR	MPH	AODV	DSR	MPH
10	44	44	45	44	43	45	44	45	46
20	47	48	48	48	47	48	47	48	48
30	47	48	48	47	47	48	48	48	48
40	46	46	47	47	45	47	46	46	47
50	43	43	46	41	41	45	39	40	45
60	43	43	46	41	41	45	39	40	45
70	44	44	47	43	43	47	42	42	47
80	46	47	48	47	47	47	44	45	47
90	47	47	48	48	47	48	46	47	48
100	48	48	48	48	48	48	47	48	48

These scenarios depend on the location of the coordinator node and the jammer node. In the first configuration, when the coordinator node is located at one corner of the topology, every node in the network forwards and sends traffic to the coordinator. Routes are made longer here, and there is interference in the middle of the topology (jammer node). In the second configuration, the non-uniform random distribution, the coordinator node is located in the center of the topology, and the remaining nodes are distributed along the area. Sixty percent of the nodes, including the jammer node, are located near the central part of the topology, and the rest of the nodes, 40%, are distributed in the remaining area. In the third configuration, all nodes, the jammer and the coordinator are uniformly distributed. Analyzed metrics are tested under reactive jamming. These information and performance metrics are shown in Tables 12–14. The metrics shown here are those that have been working along the paper. These metrics are retransmissions, CSMA retries, tries from the coordinator, achievable nodes by the coordinator and energy. The measurements were made per node, *i.e.*, data reported in the tables are average numbers per node in 100 s of simulation time. Results exhibit that the MPH protocol has a better performance in all metrics with respect to the other three protocols.

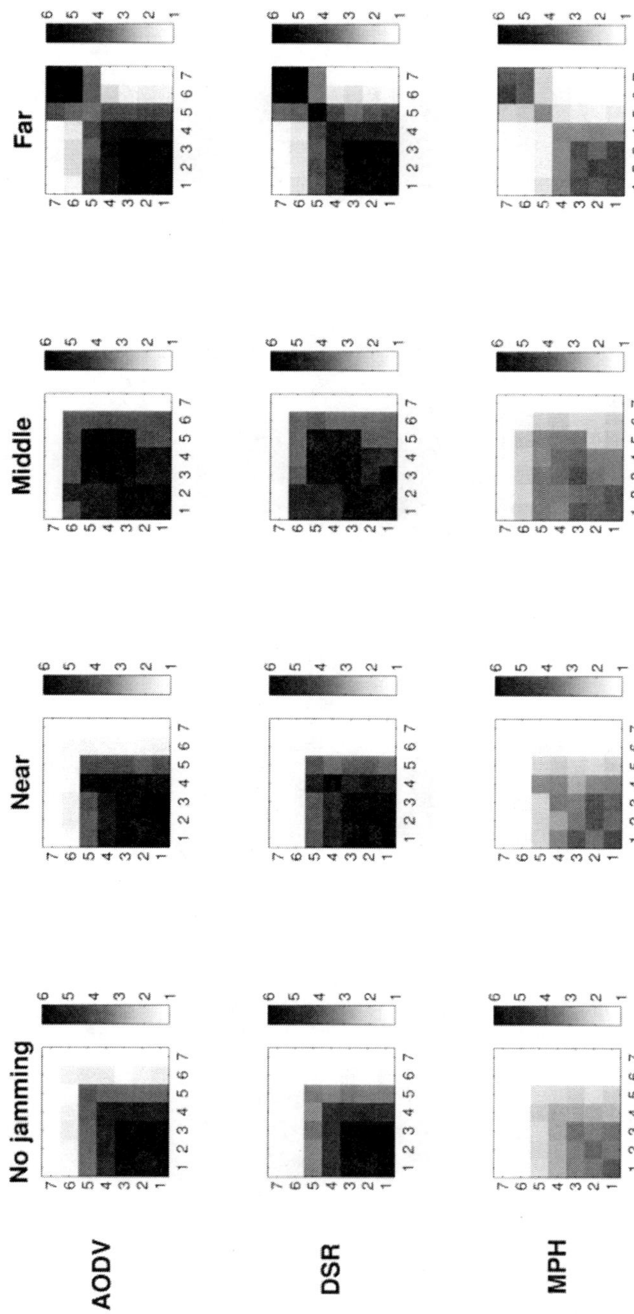

Figure 11. 2D representation of energy for the no jamming and reactive cases when the jammer node is near to, in the middle of and far from the coordinator.

Table 8. Energy model [17].

	Voltage (mV)	Current (mA)	Time (ms)	Energy (mJ)
Start-up mode	120	12	0.2	0.288
MCUrunning on 32-MHz clock	75	7.5	1.7	0.956
CSMA/CA algorithm	270	27	1.068	7.78
Switch from RX to TX	140	14	0.2	0.392
Switch from TX to RX	250	25	0.2	1.25
Radio in RX mode (processing and waiting)	250	25	4.1915	26.2
Radio in TX mode	320	23	0.58	4.26
Shut down mode	75	7.5	2.5	1.41

Table 9. Sum of values of the total energy for Figure 11.

Energy (J) for Reactive Jamming				
	No Jamming	Near	Middle	Far
AODV	149.1372	160.8547	209.6946	191.1689
DSR	136.8437	151.0130	199.6251	188.6629
MPH	86.4993	89.4854	124.3036	109.1902

Table 10. Impact of the types of jamming when the jammer node is in the middle.

Impact of the Types of Jamming of the MPH Protocol			
	Random $\rho = 50$	Random $\rho = 80$	Reactive
Retransmissions			
Against AODV	1.6% ↑	2.0% ↓	2.2% ↓
Against DSR	1.2% ↑	9.4% ↑	12.8% ↑
CSMA retries			
Against AODV	5.6% ↑	25.3% ↑	41.4% ↑
Against DSR	4.8% ↑	22.6% ↑	35.9% ↑
Tries			
Against AODV	3.1% ↑	9.9% ↑	12.8% ↑
Against DSR	2.7% ↑	9.5% ↑	20.9% ↑
Achievable nodes by the coordinator			
Against AODV	11.1% ↑	11.4% ↑	13.8% ↑
Against DSR	11.1% ↑	13.6% ↑	13.8% ↑
Energy			
Against AODV	26.3% ↑	33.7% ↑	40.7% ↑
Against DSR	23.2% ↑	30.1% ↑	37.7% ↑

The MPH protocol contribution is clearly seen in this comparison of performance metrics under the jamming attack. The three described scenarios help to visualize the impact of a jammer node with different node configurations better. A concrete contribution of the MPH protocol is self-configuration (this feature is demonstrated by achievable nodes by the coordinator), which helps to create links rapidly that were broken and to avoid packet loss (this feature is demonstrated by retransmissions and CSMA retries). Besides, we obtain better performance with low energy consumption, as we can observe in the energy metric for the three scenarios.

Table 11. Parameters for the tested scenarios.

Parameter	Value
Maximum data rate	250 kbps
Discovery packet time	10 s
Transmission power	0 dBm
Receiver sensitivity	−85 dBm
Average frame length	22 Bytes
MAC layer	CSMA/CA
Maximum number of backoffs	4
Packet loss	1%
Coverage radio	50 m
Mode	Active
Jamming	Reactive
Scenario	Static
Area	150 m × 150 m

Table 12. Scenario 1.

Topology of Figure 3 Where the Coordinator is Located in One Corner					
	Retransmissions	CSMA Retries	Tries	Achievable Nodes by the Coordinator	Energy (J)
AODV	2.72	4.03	2.23	38	4.39
DSR	2.64	3.94	2.19	38	4.25
MPH	2.21	3.57	1.91	44	3.71

7.1. A Grid Where the Coordinator Is Located in One Corner and the Jammer is in the Middle

Table 12 shows the results for a grid of 49 nodes where the coordinator is located in a corner of the grid and all nodes generate and send traffic to it.

Table 13. Scenario 2.

	Non-Uniform Random Distribution				
	Retransmissions	CSMA Retries	Tries	Achievable Nodes by the Coordinator	Energy (J)
AODV	2.91	4.42	3.14	35	4.86
DSR	2.78	4.31	3.02	36	4.56
MPH	2.42	3.82	2.19	42	4.05

Table 14. Scenario 3.

	Uniform Random Distribution				
	Retransmissions	CSMA Retries	Tries	Achievable Nodes by the Coordinator	Energy (J)
AODV	2.53	3.87	2.06	39	4.13
DSR	2.49	3.66	2.01	39	4.04
MPH	2.04	3.21	1.73	46	3.37

Regarding the retransmissions, MPH has 18.7% less retransmissions per node than AODV, and it has 16.3% less than DSR. With regard to the CSMA retries, MPH shows an average reduction of 11.4% against AODV and 9.4% against DSR. Concerning the tries, MPH has 14.3% less tries than AODV and 12.8% compared to DSR. Regarding the achievable nodes by the coordinator, MPH is the most self-configuring protocol, because it has 13.6% more achievable nodes known by the coordinator than AODV and DSR. Finally, with regard to energy, MPH has energy savings of 15.5% compared with AODV and 12.7% against DSR.

7.2. Non-Uniform Random Distribution

Table 13 exposes the results for a grid of 49 nodes where the coordinator is located in the center of the topology and all nodes generate and send traffic to it. The coordinator node is the destination. Moreover, we randomly localize 60% of nodes near the central part of the topology, including the jammer node, and the remaining nodes (40%) are distributed along the whole area in a random way.

Regarding the retransmissions, MPH has 16.8% less retransmissions per node than AODV and 12.9% less than DSR. With regard to the CSMA retries, MPH exhibits an average reduction of 13.6% against AODV and 11.4% against DSR. Concerning the tries by the coordinator, MPH is

30.2% less than AODV and 27.5% compared to DSR; this metric shows that the coordinator knows almost all routes to each node for the MPH protocol. Regarding achievable nodes by the coordinator, MPH has 16.7% more achievable nodes than AODV and 14.3% more than DSR. Finally, with regard to energy, MPH has energy savings of 16.7% compared with AODV and 11.2% against DSR.

The results of this scenario show that this configuration presents the worst case. This may be due to a high density of nodes being in the area near the coordinator node, and there is also the jammer node. This interfering node causes increased number of collisions and may actually cause more breakage of links because more nodes are affected. Because of the coordinator being the final destination, the packet loss is more affected by the presence of the jammer node.

7.3. Uniform Random Distribution

Table 14 manifests the results for 49 nodes under random positions along the area. All nodes generate and send traffic to the coordinator. Regarding the retransmissions, MPH has 19.4% less retransmissions than AODV and 18.1% less than DSR. With regard to the CSMA retries, MPH presents an average reduction of 17% against AODV and 12.3% against DSR. Concerning the tries, MPH exhibits 16% less tries by the coordinator than AODV and 13.9% less than DSR. Regarding achievable nodes by the coordinator, MPH has greater self-configuration features, such as 15.2% more achievable nodes than AODV and DSR. Finally, with regard to energy, MPH has energy savings of 18.4% compared with AODV and 16.6% against DSR.

Table 15 summarizes the percentages of MPH protocol performance against AODV and DSR. This table is intended to explain the contribution of MPH in each of the metrics studied in this work. The MPH protocol remarkably out-performs with respect to AODV and DSR; this may be due to its proactive-reactive characteristics, where its mechanisms for updating shorter routes and neighbor tables with parent-child relationships provide better versatility and speedier connection configuration and topology recovery. Moreover, MPH presents greater savings in energy consumption due to the optimization of routes and packet delivery times.

Table 15. Better performance of the MPH protocol against AODV and DSR.

	Percentage of Better Performance of the MPH Protocol				
	Retransmissions	CSMA Retries	Tries	Achievable Nodes by the Coordinator	Energy (J)
			Scenario 1		
Against AODV	18.7%	11.4%	14.3%	13.6%	15.5%
Against DSR	16.3%	9.4%	12.8%	13.6%	12.7%
			Scenario 2		
Against AODV	16.8%	13.6%	30.2%	16.7%	16.7%
Against DSR	12.9%	11.4%	27.5%	14.3%	11.2%
			Scenario 3		
Against AODV	19.4%	17%	16%	15.2%	18.4%
Against DSR	18.1%	12.3%	13.9%	15.2%	16.6%

8. CONCLUSIONS

In this work, we have compared a recently proposed protocol, the multi-parent hierarchical (MPH) protocol, with two well-known protocols, the *ad hoc* on-demand distance vector (AODV) and the dynamic source routing protocol (DSR). In the simulation results, we observed that the MPH outperforms the other ones, for all metrics that we tested: the number of packet retransmissions, the CSMA inner loop retries, the number of nodes answering the queries from the coordinator (sink) node and the energy consumption.

The assessed metrics describe the network performance against various types of jamming. The large deviations of the metric variables for AODV and DSR when attacking conditions occur may make a later construction of a jamming detector easier. In contrast, for MPH, the changes in the metrics values are not as large as the deviations shown in AODV and DSR; this could be a desirable behavior from a performance perspective, but making the construction of a new detector more difficult; thus, further metrics could be proposed and analyzed according to specific protocols.

2D representation of the metrics distribution clearly shows that the MPH protocol can better manage the jamming impact on the network. We observed that under a jamming attack, MPH intrinsically isolates the area under attack and re-routes the traffic using the rest of the network; this is true, even if the attacker keeps moving. Therefore, WSNs with the MPH routing protocol present less performance deterioration when facing jamming activity. In MPH, the periodic updating of neighbor tables produces a positive effect to deal with the links affected by a jammer node. In addition, the combination of proactive and reactive mechanisms of MPH allow the reconfiguration of the network topology without generating excessive overhead.

Regarding AODV and DSR, which are reactive protocols, if the topology is constantly changing, the overhead increases due to the route discovery mechanism; this can generate more packet loss, and the attack becomes more aggressive and extensive, as can be seen in Figures 7–9.

From the simulation results, we can conclude that the resilience provided by routing protocols for self-configuring WSNs can help to deal with jamming activity. The network topology should adapt dynamically according to the available resources and the internal mechanisms of the underlying routing protocol. The resultant performance depends not only on the time to react against the attack, but also on the efficiency to achieve a new topology.

As future work, we suggest improving the reaction under a jamming attack. Since, at present, the MPH protocol uses a persistence counter to score neighbor steadiness, the use of a larger history of the neighbor activity may reduce the number of false positives in the decision to remove a node from the neighbor tables; however, this would produce an increased delay in the reconfiguration of the topology. Therefore, by helping this mechanism with the strength of the signal and other MAC and network layer parameters, a multi-feature detector or a routing protocol with better capabilities to deal with the jamming attacks could be constructed.

ACKNOWLEDGEMENTS

This work has been partially supported by the Consejo Nacional de Ciencia y Tecnología (CONACyT, Mexico) and Tecnológico de Monterrey.

AUTHOR CONTRIBUTIONS

Carolina Del Valle built the simulator, prepared and executed the simulations, interpreted and analyzed the results and drafted the manuscript. Carlos Mex supervised the research and the methodology; he also collaborated on drafting the manuscript. Raul Monroy was involved in the design of the methodology, analysis of the results and in the manuscript review. Juan Nolazco participated in the statistical analysis and validation, as well as the manuscript review.

CONFLICTS OF INTEREST

The authors declare no conflict of interest.

REFERENCES

1. Zanella, A.; Bui, N.; Castellani, A.; Vangelista, L.; Zorzi, M. Internet of Things for Smart Cities. *IEEE Internet Things J.***2014**, *1*, 22–32.

2. Pelechrinis, K.; Iliofotou, M.; Krishnamurthy, S. Denial of Service Attacks in Wireless Networks: The Case of Jammers.*IEEE Commun. Surveys Tutor.* **2011**, *13*, 245–257.

3. Friginal, J.; de Andres, D.; Ruiz, J.; Martinez, M. A survey of evaluation platforms for ad hoc routing protocols: A resilience perspective. *Comput. Netw.* **2014**, *75*, 395–413.

4. Xu, W.; Ma, K.; Trappe, W.; Zhang, Y. Jamming Sensor Networks: Attack and Defense Strategies. *IEEE Netw.* **2006**, *20*, 41–47.

5. Del-Valle-Soto, C.; Mex-Perera, C.; Olmedo, O.; Orozco-Lugo, A.; Galván-Tejada, G.; Lara, M. On the MAC/Network/Energy Performance Evaluation of Wireless Sensor Networks: Contrasting MPH, AODV, DSR and ZTR Routing Protocols. *Sensors* **2014**, *14*, 22811–22847.

6. Del-Valle-Soto, C.; Mex-Perera, C.; Olmedo, O.; Orozco-Lugo, A.; Galván-Tejada, G.; Lara, M. An efficient Multi-Parent Hierarchical Routing Protocol for WSNs. Proceedings of the Wireless Telecommunications Symposium (WTS), Washington, DC, USA, 9–11 April 2014; pp. 1–8.

7. Kiwior, D.; Lam, L. Routing Protocol Performance over Intermittent Links. Proceedings of the Military Communications Conference (MILCOM), Orlando, FL, USA, 29–31 October 2007; pp. 1–8.

8. Mingyan, L.; Koutsopoulos, I.; Poovendran, R. Optimal Jamming Attack Strategies and Network Defense Policies in Wireless Sensor Networks. *IEEE Trans. Mob. Comput.* **2010**, *9*, 1119–1133.

9. Misra, S.; Singh, R.; Mohan, S.V.R. Geomorphic zonalisation of wireless sensor networks based on prevalent jamming effects. *IET Commun.* **2011**, *5*, 1732–1743.

10. Yu, M.; Su, W.; Kosinski, J.; Zhou, M. A new approach to detect radio jammings in wireless networks. Proceedings of the IEEE. International Conference on Networking, Sensing, and Control (ICNSC), Chicago, IL, USA, 11–13 April 2010.

11. Jiang, S.; Xue, Y. Providing survivability against jamming attack for multi-radio multi-channel wireless mesh networks. *J. Netw. Comput. Appl.* **2011**, *34*, 443–454.

12. Awerbuch, B.; Richa, A.; Scheideler, C. A jamming-resistant MAC protocol for single-hop wireless networks. Proceedings of the Twenty-Seventh ACM Symposium on Principles of Distributed Computing, Toronto, ON, Canada, 18–21 August 2008; pp. 45–54.

13. Law, Y.W.; Palaniswami, M.; van Hoesel, L.; Doumen, J.; Hartel, P.; Havinga, P. Energy-efficient link-layer jamming attacks against wireless sensor network MAC protocols. *ACM Trans. Sens. Netw. (TOSN)* **2009**, *5*.

14. Xu, W.; Trappe, W.; Zhang, Y.; Wood, T. The feasibility of launching and detecting jamming attacks in wireless networks. Proceedings of the 6th ACM International Symposium on Mobile Ad Hoc Networking and Computing, Urbana-Champaign, IL, USA, 25–27 May 2005; pp. 46–57.

15. Perkins, C.; Royer, E. Ad-hoc on-demand distance vector routing. Proceedings of the IEEE. Workshop on Mobile Computing Systems and Applications, WMCSA, New Orleans, LA, USA, 25–26 February 1999; pp. 90–100.

16. Maltz, D.; Broch, J.; Jetcheva, J.; Johnson, D. The effects of on-demand behavior in routing protocols for multihop wireless ad hoc networks. *IEEE J. Sel. Areas Commun.* **1999**, *17*, 1439–1453.

17. Texas Instruments. Application Note AN079. Available online: www.ti.com/lit/an/swra292/swra292.pdf (accessed on 10 August 2014).

18. Instruments, Texas, *A True System-on-Chip Solution for 2.4-GHz IEEE 802.15.4 and ZigBee Applications*; CC2530 Datasheet; Texas Instruments Incorporated: New York, NY, USA; February; 2011.

19. *Wireless Medium Access Control (MAC) and Physical Layer (PHY) Specifications for Low-Rate Wireless Personal Area Networks (WPANs)*; IEEE Std 802.15.4-2006 (Revision of IEEE Std 802.15.4-2003); LAN/MAN Standards Committee of the IEEE Computer Society: New York, NY, USA; 10; April; 2007; pp. 1–305.

20. Montgomery, D.C.; Runger, G.C. *Applied Statistics and Probability for Engineers*, 4th ed.; John Wiley & Sons: New York, NY, USA, 2007.

9

DYNAMIC RECONFIGURATION OF SECURITY POLICIES IN WIRELESS SENSOR NETWORKS

Mónica Pinto *, Nadia Gámez, Lidia Fuentes, Mercedes Amor, José Miguel Horcas and Inmaculada Ayala

Departamento de Lenguajes y Ciencias de la Computación, University of Málaga, Andalucía Tech, Málaga 29071, Spain

ABSTRACT

Providing security and privacy to wireless sensor nodes (WSNs) is very challenging, due to the heterogeneity of sensor nodes and their limited capabilities in terms of energy, processing power and memory. The applications for these systems run in a myriad of sensors with different low-level programming abstractions, limited capabilities and different routing protocols. This means that applications for WSNs need mechanisms for self-adaptation and for self-protection based on the dynamic adaptation of the algorithms used to provide security. Dynamic software product lines (DSPLs) allow managing both variability and dynamic software adaptation, so they can be considered a key technology in successfully developing self-protected WSN applications. In this paper, we propose a self-protection solution for WSNs based on the combination of the INTER-TRUST security framework (a solution for the dynamic negotiation and deployment of security policies) and the FamiWare middleware (a DSPL approach to automatically configure and reconfigure instances of a middleware for WSNs). We evaluate our approach using a case study from the intelligent transportation system domain.

KEYWORDS

FamiWare; INTER-TRUST; self-protection; self-adaptation; security frameworks; dynamic software product lines

1. INTRODUCTION

Recent advances in wireless communication technologies and applications have enabled the large-scale deployment of wireless sensor networks (WSNs). These networks have applications in many important and critical areas, such as military, critical infrastructure monitoring, traffic control, healthcare, environment monitoring and manufacturing [1]. Security and privacy are critical issues for most of these sensor network applications. However, providing security and privacy to WSNs is very challenging due to the heterogeneity of sensor nodes and their limited capabilities in terms of energy, processing power and memory. Thus, generic security solutions do not fit well with the special requirements of WSNs. This problem can be tackled from two different points of view.

On the one hand, there is currently an important effort being made to develop specific algorithms for WSNs in the fields of trust, security and privacy [1], such as new algorithms for authentication [2], positioning [3], network intrusion detection [4], efficient key management for dynamic WSNs [5], *etc.* Since these algorithms need to run in a myriad of sensors with different low-level programming abstractions, limited capabilities, different routing protocols, *etc.*, new algorithms are continuously appearing that improve upon previous ones. This means that sensor network applications are currently in need of being self-adapting to these changes with the lowest impact for users and without stopping the devices. Additionally, these applications need to react to changes in the environment by changing the security policies that are currently deployed in the running applications. Ideally, such applications should be capable of self-protection in the sense that they could dynamically self-adapt the algorithms used to provide security.

On the other hand, we need to consider the security of WSNs also from a software engineering point of view. Software development processes should provide security solutions that take into account the high heterogeneity and variability of sensors devices and applications and also the high variability of both security requirements and solutions.

The use of advanced software engineering technologies could hide the complexity of WSNs from software developers and could also deal with the automatic adaptation of the applications when the security and privacy requirements change. Dynamic software product lines (DSPLs) [6,7] allow managing both variability and dynamic software adaptation, so they can be considered a key technology in successfully developing self-protected WSNs' applications. In this paper, we propose a self-protection solution for WSNs based on the use of DSPLs. Basically, DSPLs produce software products capable of adapting to requirements that may change at runtime, by modeling the elements that could be adapted dynamically as dynamic variation points. The runtime adaptation of security policies for heterogeneous systems implies both the modeling and the dynamic negotiation of security policies. According to the literature on security solutions for pervasive applications, there is a lack of powerful techniques to address the problem of security policy negotiation, interoperability and dynamic enforcement.

In this paper, we address all these issues by combining the INTER-TRUST (INTERoperable TRUST Assurance Infrastructure)security framework [8] and the FamiWare middleware [9]. INTER-TRUST is a dynamic and scalable framework to support trustworthy services and applications for heterogeneous networks and devices. It is mainly based on the dynamic enforcement of interoperable and changing security policies [8] by modeling secure interoperability policies with different constraints. As far as we know, INTER-TRUST is the only existing framework for the dynamic negotiation of security policies in heterogeneous environments. However, INTER-TRUST does not deal with the structural variability of WSNs, where there are many sensors of different types (such as the MICAz, a 2.4 GHz, IEEE 802.15.4 compliant mote module used for enabling low-power, wireless, sensor networks, or the Java SunSPOT, the Sun Small Programmable Object Technology) and each sensor can have its own configuration. This is one of the main contributions of FamiWare. FamiWare is a reconfigurable family of middlewares for ambient intelligence (AmI) systems, including WSNs. Using a DSPL approach, FamiWare is able to automatically configure and reconfigure instances of the middleware for all of the nodes of a WSN. Thus, FamiWare solves the problems of modeling and reasoning about WSN heterogeneity and variability. In our approach, we combine both approaches to enable the generation of security solutions for heterogeneous sensor networks, where: (1) the security can be negotiated at runtime (capability provided by INTER-TRUST); (2) the security variability is formally modeled (capability provided by FamiWare); (3) the system is reconfigured to enforce the new security policies dynamically at runtime (capability provided by FamiWare and INTER-TRUST); and (4) the system reconfiguration is customized for each heterogeneous device in the AmI system (capability provided by FamiWare).

We will show how the INTER-TRUST framework can be instantiated into an instance of the FamiWare middleware generated for WSNs using a contextual advisory speed application for intelligent transport systems (CAS/ITS). The rest of the paper is organized as follows. After this Introduction, Section 2 introduces the background information required to understand our approach and the case study that we are going to use. Then, Section 3 presents the main challenges that need to be solved so as to provide security in WSNs and briefly describes our approach to address these challenges. Sections 4 and 5 present the details of our approach, including the steps required to integrate INTER-TRUST and FamiWare and the dynamic reconfiguration process in FamiWare. The evaluation of our approach is discussed in Section 6. Finally, Section 7 compares our approach with related work, and Section 8 presents our conclusions and future work.

2. BACKGROUND

This section introduces the background required to understand the approach presented in this paper. It is basically about dynamic software product lines and dynamic security frameworks. We focus on how these techniques are used in FamiWare and INTER-TRUST, respectively. Additionally, we present the case study used throughout the paper.

2.1. Dynamic Software Product Lines in FamiWare

Dynamic software product lines (DSPLs) [6,7] represent an emerging field that is producing software products capable of adapting to requirements that change at runtime. A product is defined as a unique composition of features from a set of common and variable features (*i.e.*, variants), where some of the variable features may vary at runtime (*i.e.*, variation points). Therefore, the products have the capacity to reconfigure themselves by binding variation points at runtime, several times, in accordance with the changes in the environment. DSPLs use a variability model to specify the variation points that can be dynamically modified. In particular, in our approach, the use of non-standard languages or formalisms is avoided, and we use the Common Variability Language (CVL) [10] proposed as the standard by the Object Management Group (OMG) [11]. CVL is a domain-independent language for specifying and resolving variability. The CVL variability models allow modeling the variability separately from the base model, but both the variability and the base models are connected and can be managed using the same tool (see Figure 1). In CVL, VSpecs (Variation Specifications) are tree structures representing choices or elements (called "features" in most SPL terminologies) and can include logical constraints defined in a subset of the Object Constraint Language (OCL) [12]. The main characteristics of CVL that we use to model the FamiWare variability model are shown in Figure 1. In DSPLs, a variability model configuration is the set of features currently bound at runtime that represents a particular product of the SPL. In this way, dynamic reconfiguration is defined in terms of replacing the current variability model configuration with a new configuration in which the variation points have been re-bound to adapt them to a context change. Therefore, using the CVL approach, we automatically generate both the initial configuration and also the successive adapted configurations.

As shown in Figure 1, the inherent heterogeneity and the variability of WSNs are modeled in FamiWare using CVL variability models (*i.e.*, the VSpec trees). FamiWare proposes the customization of the piece of software related to the middleware platform that is deployed in each device of the WSN according to: (1) the device features (device [1..*] feature inFigure 1 with a 1..* cardinality to represent the different devices in the system); (2) the application requirements (application feature); and (3) the global characteristics of the network used to connect the different devices (network feature). Each device is represented, among other features, by its type (choice DeviceType) and the FamiWare services that need to be instantiated in that particular device according to the application's requirements and the network's characteristics (service choice). Regarding their types, devices can be of high capacity (such as desktop computers or smartphones) or they can be sensors (sensor choice). The sensors, among other things, have one or more sensor units (choice SensorUnit) (e.g., accelerometer, distance or movement sensors). For FamiWare, the variability is decided for each device, so the FamiWare services instantiated in each device can be different. Among other services, FamiWare is composed of three optional services: monitoring, context awareness and reconfiguration. The FamiWare monitoring service observes different types of contexts (e.g., memory consumption, battery level, status of network connection). These contexts are optional choices that will be selected or not in a particular configuration depending on the capacities of the sensor nodes. The context awareness service receives events with the monitored data (modeled as CVL variables in Figure 1) and checks whether these data imply that a context change has occurred (e.g., the memory consumption of a sensor is very high). When a context change is detected, the reconfiguration of FamiWare may be necessary to maintain proper functioning (e.g., modifying the tasks of certain sensors to reduce memory consumption). In order to do that, the context awareness service, by means of its product generation module, triggers the automatic election (context-based election) of the new variability model configuration adapted to the new context. This new middleware configuration is automatically generated by FamiWare (e.g., a configuration with a different security algorithm that consumes less). This is done at the model level giving as input the new context conditions and using the dependencies between features or the cross-tree constraints. The configuration that deals with these context conditions is automatically calculated by CVL. Then, a reconfiguration plan with the implementation level tasks for switching from the previous configuration to the new one is also generated by FamiWare. For that, FamiWare takes as inputs the previous running configuration and the next one, uses a difference operator (which has been previously implemented) to obtain the differences between the two configurations and generates the plan. Finally, this plan is interpreted and executed in the sensors by the reconfiguration service. Note that Figure 1 shows a reduced version of the complete FamiWare variability model, where the details of the application and the network features, among other things, have been omitted.

The FamiWare base model consists of a microkernel plus services architecture, as shown in the lower part of Figure 1. The composition between services and the application and between the services themselves is performed with a publish/subscribe event-based mechanism, specifically a reduced implementation of the data distribution service (DDS) [13], which is an OMG standard for publish/subscribe middlewares.

2.2. Dynamic Security Frameworks: INTER-TRUST

A security framework is a reusable set of modules that can be customized to develop trustworthy applications. The framework describes the interfaces and main functionality of each module, as well as the flow of control between them [14]. There are two different types of modules that can be distinguished in a security framework. On the one hand, the framework contains a set of modules that implement the security concerns, such as authentication, authorization, privacy, encryption, *etc.* On the other hand, the framework contains modules to deploy and manage the security of an application (*i.e.*, when and how the security concerns are incorporated into an application). From among these modules, the framework can incorporate specific modules to adapt the security level of the applications at runtime. In this case, we are referring to dynamic security frameworks.

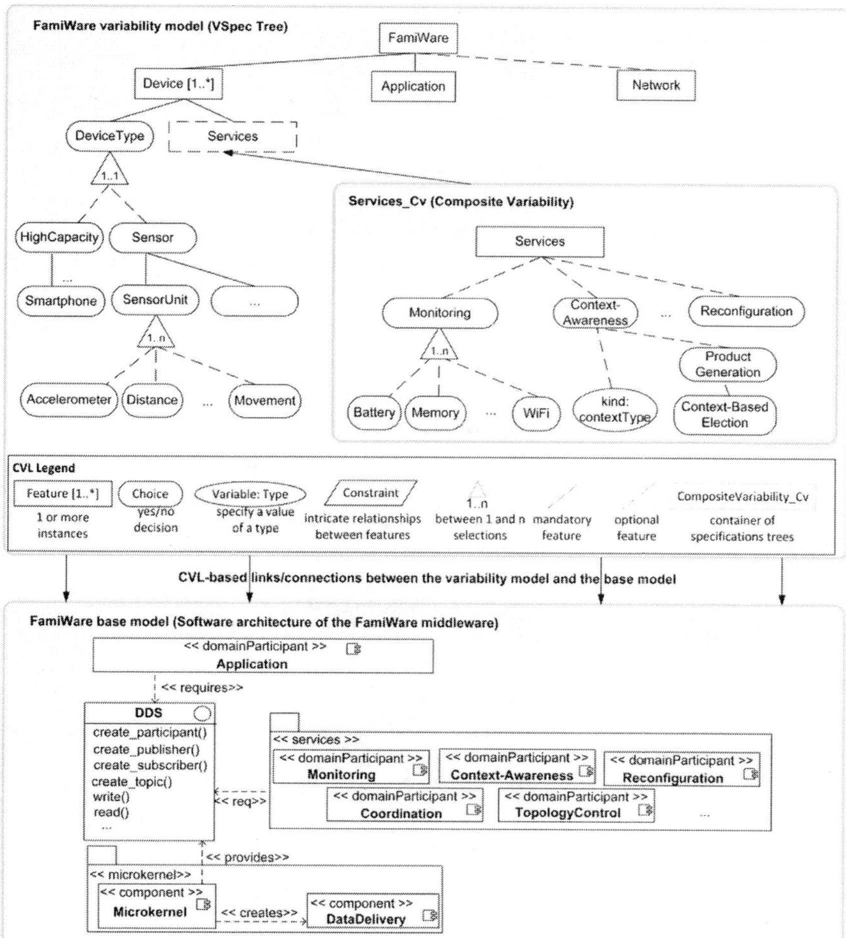

Figure 1. FamiWare variability model.

The INTER-TRUST framework is a dynamic security framework to support trustworthy applications based on the enforcement and dynamic adaptation of security policies. As shown in Figure 2, in the INTER-TRUST framework, security policies are first specified using a security editor (e.g., MotOrBac, an Organization-Based Access Control policy editor [15]) and then negotiated between the different parties in a communication using a negotiation module. When the security requirements change at runtime, the security policies are (re)negotiated. The negotiated security policy is then analyzed and interpreted by the policy engine and the policy interpreter modules. These modules are responsible for identifying changes in the security policy that require adapting the security concerns (e.g., authentication, encryption, *etc.*) deployed inside the application dynamically, at runtime. When security policies need to be dynamically deployed and/or reconfigured at runtime, the aspect generation and the aspect weaver modules are in charge of obtaining the differences between the old and the new security configurations. Additionally, these modules incorporate or eliminate the corresponding security modules in the application itself. The environment is monitored using the notification module and the Security Monitoring tool. The monitored information is used by the context awareness module to notify the changes in the context. These changes can generate a new negotiation of the security policy and the negotiation-interpretation-deployment process begins again. Changing the security at runtime can introduce security flaws into the application. In order to avoid this, different active and fuzz testing techniques (Fuzz Testing and Active Testing tools) are provided as part of the framework in order to confirm that the dynamic changes are not harmful.

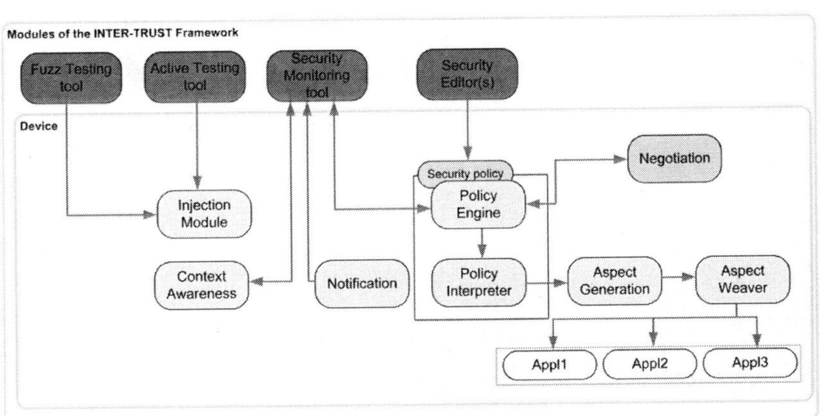

Figure 2. The INTER-TRUST framework.

2.3. Case Study: Intelligent Transportation Systems

One of the most relevant applications today, where the security issues in WSNs play a very important role, are the intelligent transportation systems (ITS) [16]. These systems have the potential to improve mobility, safety and security, while ensuring energy efficiency and reducing the environmental impact of transportation systems. These systems are composed of different kinds of devices installed at several locations, such as inside the vehicles, on the road or in the traffic signals. For instance, the vehicles can be equipped with several sensors with different sensor units, such as acoustic, accelerometers, temperature, the vehicles' on-board unit (OBU) computers and GPS devices. These sensors are used to ensure safe driving, being responsible, for instance, for measuring the distance to obstacles around them. The vehicles also use information received from surrounding vehicles, such as their speed or distance, and are able to communicate with each other in order to cooperate for safety and other reasons. Furthermore, the traffic signals send information to the vehicles informing them of speed limits and general traffic regulations. ITSs have a wide variety of interesting applications, such as road safety applications (accident warnings, red-light warnings, road congestion, contextual speed advisory, *etc.*), information dissemination applications (parking spots, traffic warnings/conditions, fuel prices, landmarks, *etc.*), entertainment applications (file sharing, advertisements, voice communication with vehicles nearby, *etc.*) or assisted driving. Additionally, vehicles can be considered as distributed sensors that can collect environmental data (such as average speed, potholes, temperature, pollution, *etc.*) and send a report to base stations. This infrastructure composed of the vehicles' sensors, GPS and OBUs, road sensor units (RSUs) and other sensors or actuators installed in traffic signals, is common to all of these applications. Similarly, some of the requirements, such as coordination, monitoring or security, are also common to all of the applications.

In particular, the security needs of all of these applications are quite important, especially when dealing with safety-related services. In these cases, to know whether a node in the network is being uncooperative or malicious is really crucial to prevent intentionally wrong instructions being sent to a vehicle. Furthermore, these systems are very dynamic, and their contextual environment changes very frequently, so they need different security policies adapted to the changing situations. Additionally, the ITS infrastructure is composed of extremely heterogeneous devices where both the monitoring and the deployment and management of the security policies must be performed in different ways.

In this paper, we illustrate these needs with a specific ITS application, a contextual speed advisor (CSA) application [16], where vehicles are informed about the speed they have to follow, taking into account the contextual environment. As shown in the scenario in Figure 3, this kind of application has the same common infrastructure as previously described, with sensors, OBUs, base stations, and so on. In this application, basically the base stations periodically receive the information from the sensors and other devices (e.g., the location of cars, information about crashes, traffic congestion or weather conditions). With this information, the base stations calculate the recommended speed for each vehicle on a certain stretch of road and send it to the vehicles (through the RSUs to the vehicle OBUs or to other user devices, such as smartphones).

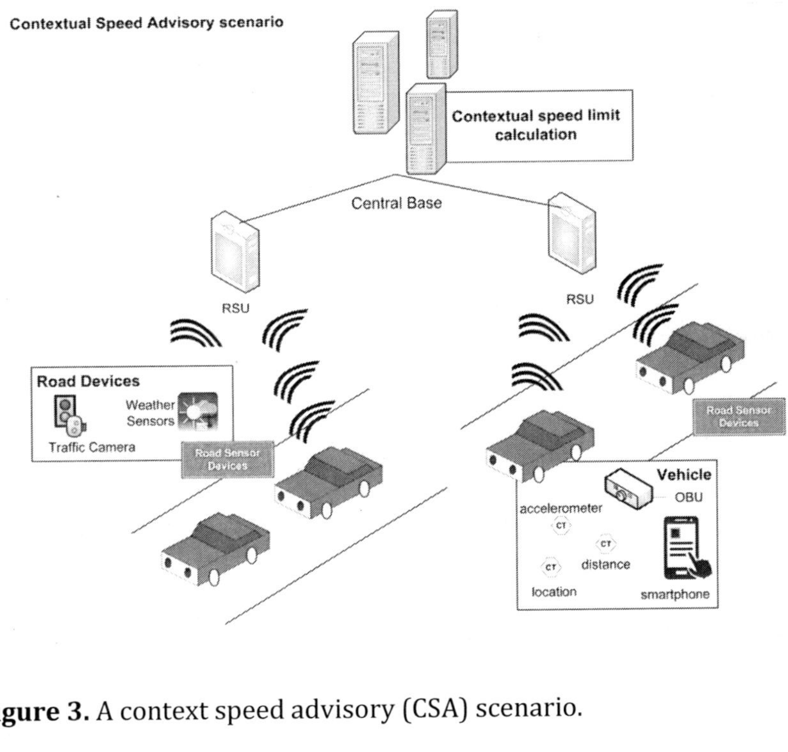

Figure 3. A context speed advisory (CSA) scenario.

In the CSA scenario, avoiding malicious attacks is extremely important. For example, to indicate that the recommended speed is 100 kph when you have traffic congestion in the vicinity could cause an accident. Therefore, the messages sent between the nodes and the base station must be properly encrypted, and all of the nodes must be trustworthy. Therefore, the self-protection of the whole system is required. However, very hard security policies have a penalty in efficiency, especially in energy and performance, two typical requirements of these applications. Therefore, the security policy to be applied in each situation must be negotiated to make a trade-off between the necessary security and the efficiency. Additionally, as these situations can change quickly, these policies have to be (re)negotiated. For instance, in the case where malicious software is suspected in a given WSN (detected by intruder detection algorithms [17]), the security needs to be increased, and the suspect nodes must be isolated. To the contrary, on roads where the main interest is to preserve the lifetime of the system and there is no risk of malicious attacks, the security can be a little bit weaker. Similarly, in situations where communication performance is required (e.g., high traffic density), the security could also be relaxed. On the other hand, in exceptional circumstances, the interception of data sent by certain vehicles (e.g., a presidential motorcade) has to be totally avoided. In this case, the security policy applied to these specific nodes must be modified. Furthermore, other special situations, such as emergencies or attempted terrorist attacks, require another kind of global security and/or intrusion detection.

3. CHALLENGES AND OUR APPROACH

In this section we describe, in more detail, the main challenges that need to be taken into account in order to add security to WSNs and to dynamically (re)configure the security level at runtime. We also indicate how our approach copes with these challenges.

Challenge1: Build secure applications that run on heterogeneous nodes with limited capabilities. The nodes of the WSNs are heterogeneous, and thus, the security solution deployed at each node needs to be adapted to the characteristics of every node. Furthermore, these nodes have limited capabilities, and the security solutions need to be carefully specified to avoid consuming too many resources. This means that we need to define a family of security solutions, providing different implementations for the different nodes in the WSN, while also taking into account the constraints on the resources available for each node. Our approach: Using the software product line mechanisms, already existing in the development process defined as part of FamiWare, we will be able to generate customized instances of the FamiWare middleware, each one ready to be run on a specific node of the WSN. In this way, every instance of FamiWare has just the security services or algorithms required by the applications installed in each node. Therefore, we avoid loading a big chunk of security code into the limited tiny devices. We will integrate the new security components defined by INTER-TRUST as part of the FamiWare product line architecture.

Challenge2: Dynamic negotiation of security policies for WSNs. The conditions of the running environment of the WSNs applications could change. Then, the security policies may need to be adapted to the new conditions. This adaptation requires a previous negotiation between the different parts of the application. Our approach: As previously said, INTER-TRUST is the only framework that provides dynamic negotiation of security policies. By integrating the INTER-TRUST framework with FamiWare, we provide a middleware that allows the dynamic negotiation of the security policies for the sensor nodes already included as part of FamiWare.

Challenge3: Monitor the changes in the context. These changes can be general (e.g., changes in the amount of memory or the battery level available) or can be specific to security. In order to react to these changes, the environment needs to be monitored. Our approach: Both INTER-TRUST framework and FamiWare middleware include specific components for monitoring the environment. Therefore, we will add those security components provided by INTER-TRUST that are not present in FamiWare. The main contribution of the process proposed here is to guarantee that among all of the existing monitoring components, only the most necessary and appropriate ones are incorporated into the solution for each sensor node. This is done via the configuration process provided by FamiWare.

Challenge4: Define a dynamic reconfiguration service to endow applications with self-protection. When the environment or the security requirements change at runtime, the system must be self-protected to maintain a proper and secure functioning. Our approach: Firstly, the new security policies to be used need to be negotiated using INTER-TRUST. Secondly, the new security policy needs to be analyzed to determine whether the changes to the security policy imply changes in the system's components. If changes in the system's components are required, then a reconfiguration plan needs to be generated to carry out these changes without stopping or crashing the system. The last two tasks are performed by FamiWare. These plans contain the implementation level activities to be carried out in order to instantiate the new security policies.

Challenge5: Provide an efficient solution able to run on a WSN. As WSNs are composed of a large number of tiny devices, the self-protection mechanism should consume the least number of resources as possible. Our approach: The FamiWare reconfiguration mechanism pays special attention to efficiency, as we will show in the evaluation (see Section 6). We focus on proving that resource consumption of this mechanism is optimal for a typical sensor type (MICAz). We also check that our solution is scalable to the number of nodes. Furthermore, we demonstrate that the overhead and latency of the self-adaptation is not critical compared with the benefits of having secure applications in WSNs.

4. INTEGRATION OF FAMIWARE AND INTER-TRUST

As explained, FamiWare is a family of middlewares specially designed and implemented for dealing with the heterogeneity of WSNs devices. It also provides a mechanism to be reconfigured at runtime when the context changes. However, it does not focus on providing security. Therefore, if we want to extend FamiWare with the benefits provided by the INTER-TRUST framework, the first thing to do is to integrate INTER-TRUST into the FamiWare family. As shown inFigure 4, this implies several tasks that have been organized into two main steps. The first step consists of the integration of the INTER-TRUST modules and the INTER-TRUST security concerns into the FamiWare product line. This implies the extension of both the variability model and the base model of FamiWare. This step allows the generation of different FamiWare middleware configurations, customized for the necessities of each device, including the customization of the security modules and concerns. In the second step, the security policies specified in INTER-TRUST and the security concerns (to be deployed for enforcing those security policies) are taken as input to define and implement the reconfiguration actions. These reconfiguration actions need to be executed in order to generate a new configuration adapted to the new security context. Therefore, these actions must be part of the reconfiguration plans that FamiWare generates at the implementation level to switch from the current FamiWare configuration to the next one. This step enables the security requirements to be adapted at runtime.

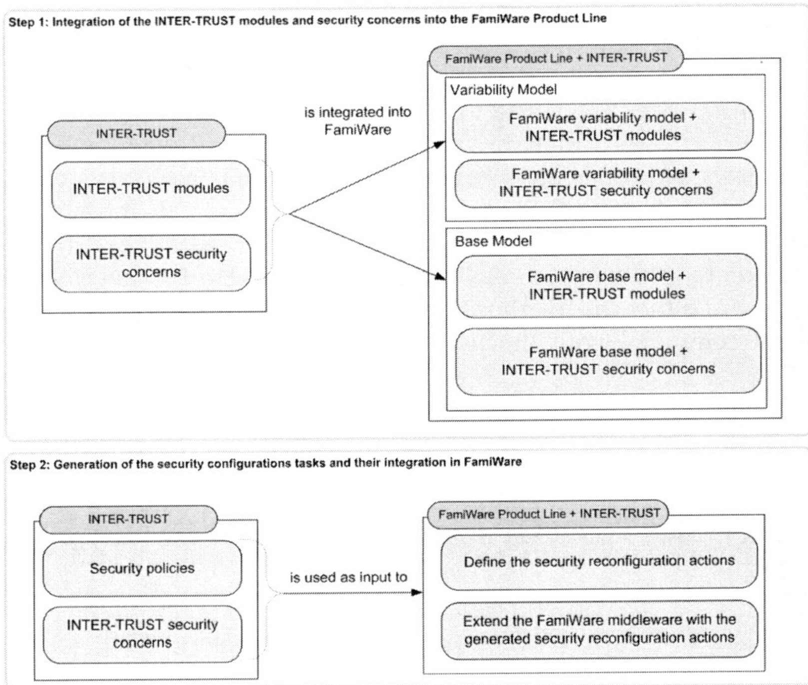

Figure 4. Steps for the integration of INTER-TRUST into the FamiWare family of middlewares.

In the rest of this section, we explain these steps in more detail. The first step is detailed in Section 4.1 and the second step in Section 4.2. Finally, Section 4.3 shows different configurations of the FamiWare middleware with security for different devices of our ITS case study.

4.1. Integration of the INTER-TRUST Modules and Security Concerns

This section is organized in two main subsections. Section 4.1.1 principally focuses on the integration of the INTER-TRUST modules, while Section 4.1.2 focuses on the integration of the INTER-TRUST security concerns.

4.1.1. Integration of the INTER-TRUST Modules

In Section 2.1, we described how a different configuration of the FamiWare middleware could be deployed in each device according to that device's characteristics, the applications that this device runs and the network used to connect several devices. Now, with the integration of the INTER-TRUST framework into the FamiWare middleware, the security also has to be taken into account to configure the FamiWare middleware. To do so, as shown in Figure 5, the VSpec tree of FamiWare is extended with the security feature. Due to the correspondence between feature models [18] and CVL VSpec trees, for simplicity, we use the term 'feature' throughout the paper to refer to the term 'VSpec tree node'. Since the security concerns instantiated could be different for each device, the security feature has been included as a child of the device feature instead of defining it as a child of the FamiWare feature. This feature is a CVL composite variability. This means that it is a container of another VSpec tree describing the variability of all of the security concerns that could be instantiated in each device (this is described in Section 4.1.2).

In addition to the security concerns, the INTER-TRUST modules also need to be incorporated as part of the services provided by FamiWare. Thus, we need to extend these services for the special requirements of the INTER-TRUST dynamic negotiation of security policies. On the one hand, FamiWare does not provide any service for negotiation. Thus, we need to extend the variability model of FamiWare with a completely new service, the negotiation service, which is included at the same level as the monitoring, context awareness and reconfiguration services. On the other hand, the already existing FamiWare services need to be extended to include the monitoring, context awareness and reconfiguration features related to security that are specific to INTER-TRUST:

- Monitoring: This is represented in the FamiWare VSpec tree as an optional element with one child or more for each context that can be monitored by the middleware. Therefore, as we need to monitor the issues related to security (e.g., vulnerabilities, security algorithm overheads), we have to add a new element to the children of the monitoring element called security monitoring in Figure 5. This variable element has two children, the Montimage Monitoring Tool and the notification features. They are described in Section 2.2 and are the modules and tools used in the INTER-TRUST framework to monitor the application's security.

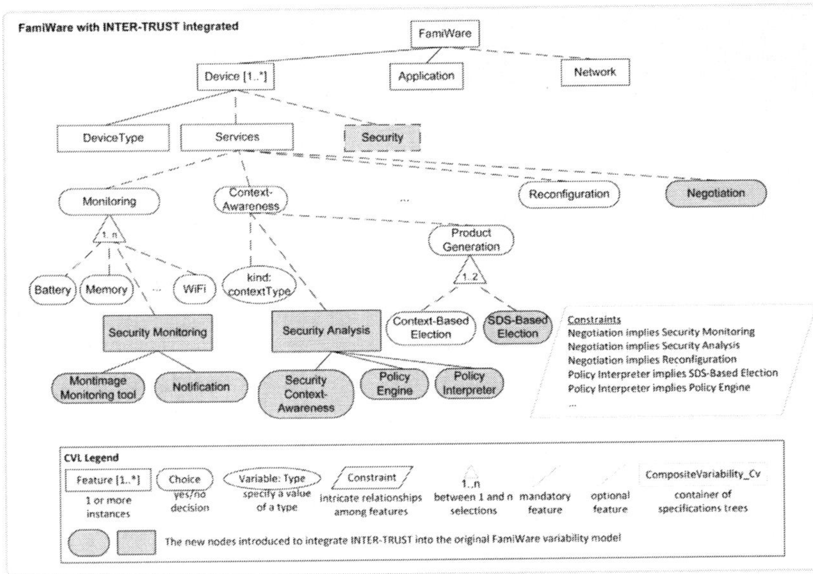

Figure 5. FamiWare + INTER-TRUST variability model.

- Context awareness: The context awareness feature of FamiWare is represented by two elements. One is the kind of context of which a certain application is required to take care. The other is the list of changes to be executed to generate a new product when a context modification implies the system's reconfiguration. In INTER-TRUST, this is carried out by the security context awareness together with the policy interpreter and the policy engine modules. These modules decide whether or not a new security policy must be negotiated and the changes to be performed in the system security after the negotiation. Thus, a new feature called security analysis has been added as a child of the context awareness feature of FamiWare that includes the security context awareness, policy engine and policy interpreter children. Moreover, a new feature has been added as a child of the product generation feature to indicate that, in the case of security, the new product configuration needs to be generated, taking into account the information dynamically generated by the policy interpreter module of the INTER-TRUST framework. Thus, now, the reconfiguration plans could be generated using different processes: the context-based election feature represents the original process used by FamiWare; the Security Deployment Specification (SDS)-based election feature represents the process that generates the reconfiguration plans taking as input the information generated by the policy interpreter module of INTER-TRUST. The SDS [19] is a custom format defined by INTER-TRUST to indicate the security concerns that need to be deployed for each security policy and the security functionalities that need to be provided by each concern (e.g., the authentication concern with a user-password functionality). In both cases, a reconfiguration plan is generated at runtime by FamiWare.

- Reconfiguration: In FamiWare, this service reconfigures the nodes of the WSN by executing the corresponding reconfiguration plan in each node of the network. With the integration of INTER-TRUST, this module is also in charge of reconfiguring the security of the system by instantiating the new negotiated security policy in all of the necessary sensor nodes of the network. Since the context awareness module generates an reconfiguration plan with the same format for both generic and security-specific reconfigurations, the original reconfiguration service of FamiWare does not need to be extended and can be used as is to reconfigure the security of the FamiWare applications. However, the reconfiguration tasks for security need to be defined and implemented as explained in the following subsection.

The main advantage of modeling the security concerns and the modules of INTER-TRUST as variable characteristics of the FamiWare product line is that it is possible to generate different middleware configurations with different security policies for the heterogeneous devices of a WSN. Furthermore, using the variability model at runtime, it is possible to renegotiate these security policies at runtime.

As said, in CVL, the OCL constraints specify the dependencies between the aforementioned features. In addition to the constraints already specified for the FamiWare variability model, Figure 5 shows part of the constraints for the new features. For instance, the constraints of the figure specify that the selection of the negotiation feature implies the selection of the security monitoring and the security analysis features or that the selection of the policy interpreter feature implies the selection of the SDS-based election feature.

Note that we have focused on showing how the variability model of FamiWare is extended because we consider that this is the most interesting part of Step 1 shown in Figure 4. The base model is also modified by adapting the implementation of the INTER-TRUST modules to the event-based component model of FamiWare and by including the new components in the software architecture of FamiWare

4.1.2. Integration of the Security Concerns

Figure 6 shows the VSpec tree that models the variability of the INTER-TRUST security concerns (*i.e.*, the VSpec tree contained in the security composite variability shown in Figure 5). The top of Figure 6 shows the abstract part in which all of the security concerns, functionalities, attributes and parameters that can be reconfigured at runtime are specified (*i.e.*, the variability specification tree for security). The bottom of Figure 6 shows a representation of the specific implementation of the security functionalities encapsulated in the FamiWare components (*i.e.*, the base model of the security concerns). Details of the parameters of each security concern in the variability specifications tree and in the FamiWare components are hidden in Figure 6 to simplify it. Some of these details are shown in Figure 7 for the authentication functionality and are detailed below. Finally, the security features of the tree and the specific functionality of the FamiWare components for security are linked by using the CVL variation points (middle of Figure 6). A particular selection of the features in the tree defines a particular configuration of the security components.

As stated, Figure 7 shows the kinds of details that were omitted for the security concerns in Figure 6. In particular, the required configuration for the attributes and parameters of the authentication concern are shown. The graphical elements shown in grey indicate those elements selected in a particular configuration. In this case, the VSpec NetworkAuthhas been selected to indicate that a dynamic authentication and key establishment protocol [5] will be used as the authentication mechanism. Moreover, the configuration also includes the parameters for the certificate authority (TrustedCA) such as the information about the organization that issued the certificate and the parameters for the key caches that will store the session keys to be used with the certificate (nodeID, Keyand KeyLifetime). Note that this is only an example, and the variability model allows different instances of each component modeling a security concern to be created and configured. Moreover, as illustrated in Figure 8 in Section 4.3, the variability model also allows the use of different authentication mechanisms for different parts of the same application.

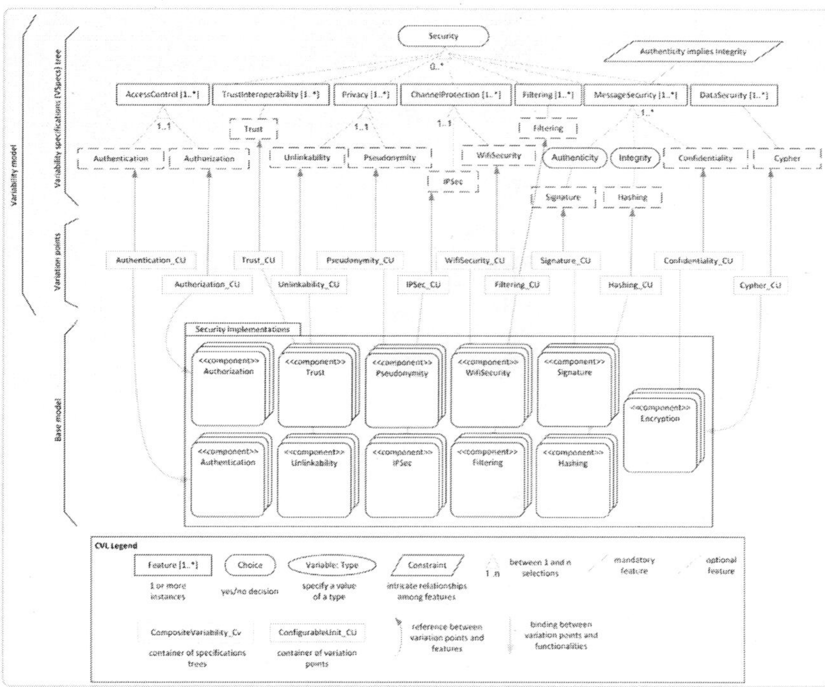

Figure 6. Composite variability: security concerns.

Finally, the FamiWare base model is extended to integrate the security aspects provided by INTER-TRUST. Security concerns are implemented in INTER-TRUST using the aspect-oriented programming approach [20]. However, the approach presented in this paper is not affected by the use or not of aspects. FamiWare components communicate using a publish/subscribe interaction model. This means that in order to integrate the security aspects provided by INTER-TRUST, we need to define a FamiWare wrapper component that wraps the behavior provided by the advice of the INTER-TRUST aspects. This is straightforward, since the aspects in INTER-TRUST act also as a wrapper of the security behavior, which is implemented independently of a particular technology.

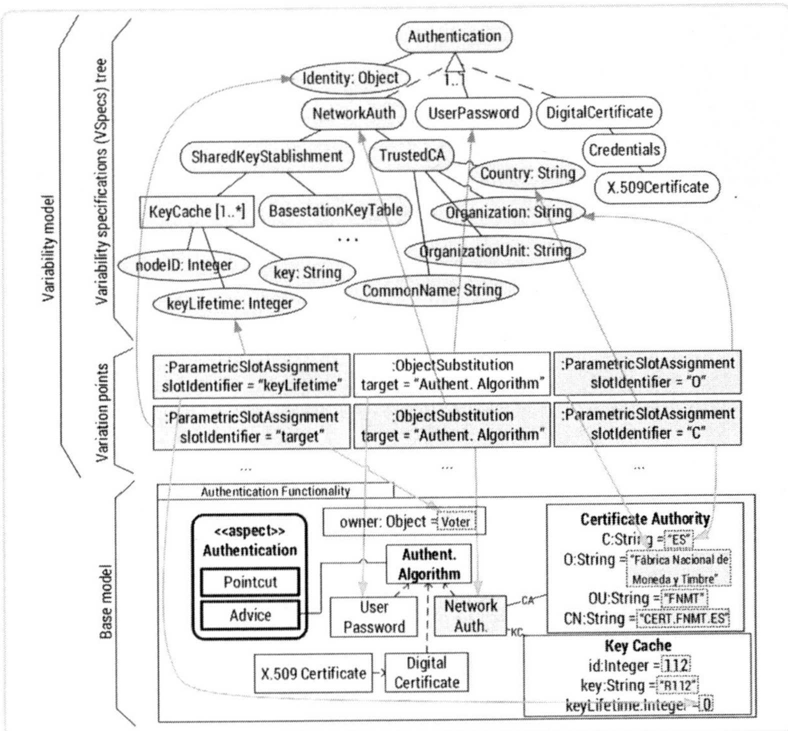

Figure 7. Resolution model of the authentication concern.

4.2. Generation and Integration of the Security Reconfiguration Actions into FamiWare

In order to maintain the correspondence between the running code and the model, FamiWare defines a two-level mapping between the modifications of the features in the VSpec tree and the implementation level tasks [21] to be executed for the system reconfiguration.

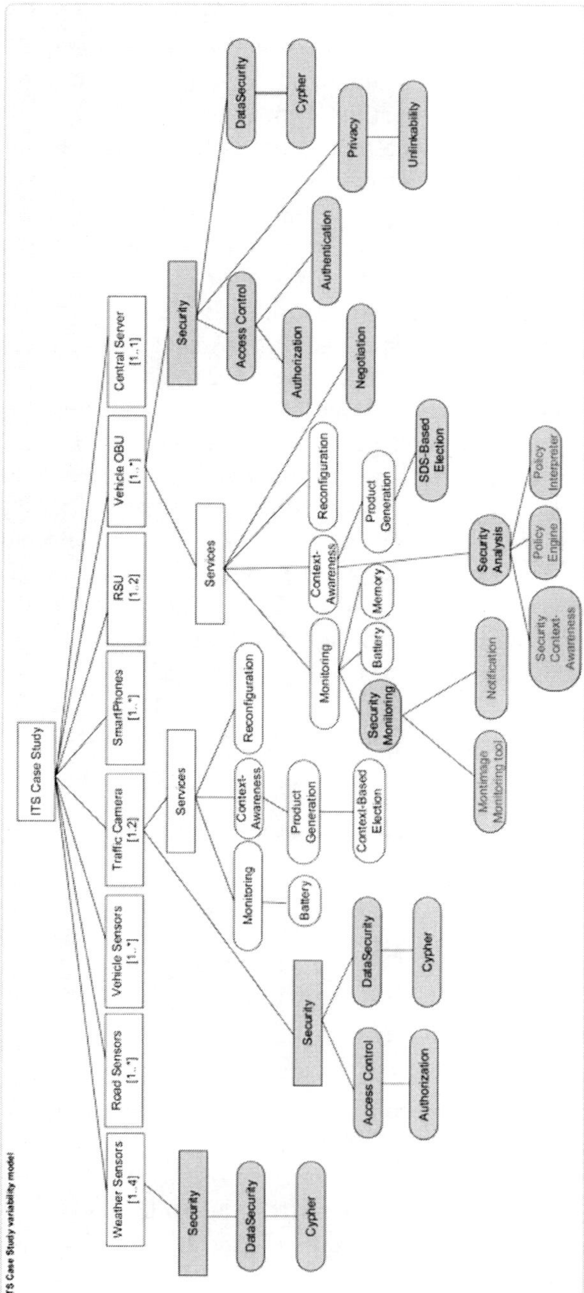

Figure 8. Intelligent transport system (ITS) case study variability model.

The first mapping is a table that specifies for each feature of the VSpec tree how the presence/absence of that feature implies the execution of a set of generic reconfiguration tasks. Examples of these generic tasks are activating or deactivating a sensor, changing the role played by a sensor, adding a service to a certain FamiWare middleware instantiation, removing a running service or changing the configuration of a service. Since these tasks are generic, they are already provided by FamiWare and are directly reusable in the extension of FamiWare with INTER-TRUST. An example of this correspondence would be: "when the cypher feature was not selected in the running product and is now selected in the VSpec tree representing the adapted product, a reconfiguration task to add the service for encrypt/decrypt messages has to be executed". At the implementation level, this corresponds to the task *AddService*($Device_n$, $Service_m$) with two input parameters (the device identification and the service name). Then, the tasks can have input parameters; thus, we define a task as $t(p_1, ..., p_n)$, where n is the number of parameters, which can be variable. A FamiWare reconfiguration plan is a sequence of tasks specified in OWL-S [22], an ontology of services for semantic descriptions of web services. For instance, examples of tasks, in this reduced version of OWL-S, to add the encryption service to a traffic camera and to configure it with a particular encryption algorithm, would be like this:

```
<AtomicProcess ID="AddService">
    <Input ID="TrafficCamera2"/>
    <Input ID="Encryption"/>
</AtomicProcess>
<AtomicProcess ID="ConfigureService">
    <Input ID="Encryption"/>
    <Input ID="encryptAlg2"/>
</AtomicProcess>
```

Using these generic reconfiguration tasks, FamiWare permits the reconfiguration of both the hardware of each device and its internal FamiWare architecture. Regarding hardware reconfiguration, FamiWare could, for instance, switch the radio chip of the sensor to on/off or change the whole state of the sensor. With regards to the modifications in the software architecture, it is possible to perform coarse-grained modifications, such as remove or add a service, and it is also possible to make fine-grained modifications, like changing some parameters of a service.

The second mapping specifies how these generic reconfiguration tasks are implemented for each device, application and/or new service. Since the implementation may be different for each device, application and/or service, a list of specific reconfiguration actions need to be implemented for each generic task. Some actions are specific to a concrete type of device, as is the case of putting to sleep a sensor device (*i.e.*, switching off a MICAz mote), or to a concrete application, such as the specific places in the application where a security concern needs to be ensured (e.g., authentication must be ensured when the vehicle interacts with the server for the first time).

The current implementation of FamiWare already provides the code of the device-specific actions for a broad range of sensors/devices and platforms (e.g., TinyOS, SunSPOT). However, in the case of the reconfiguration actions that are specific to a certain application or to a new service (e.g., security), the code of the particular actions to be executed for reconfiguring the systems must be manually provided to FamiWare. Thus, according to the second step shown in Figure 4, the integration of INTER-TRUST into FamiWare requires that the FamiWare variability models of Figures 5 and 6 are complemented with the list of security reconfiguration actions that need to be executed to reconfigure the application security level at runtime. Furthermore, the mapping between the generic reconfiguration tasks and these security-specific actions needs to be defined for each security concern and security policy. This means that, in order to provide security in FamiWare, we need to implement these basic actions for activating each security concern (e.g., encryption) that have been previously incorporated in the security feature of the FamiWare VSpec tree. For instance, in the case of adding the cypher service using encryption algorithm, encryptAlg2, we need an implementation action to add the component that implements the encryption algorithm. Following the publish and subscribe model of FamiWare, the code to add the service to encrypt messages in a sensor/device would be the following:

```
addComponents(<sensorID>, encryptAlg2)
publisher(<sensorID>, encrypted)
subscriber(<sensorID>, encrypted)
```

This means that, now, Traffic Camera 2 publishes/receives encrypted messages. Note that this is done once, and then, after the action has been implemented and incorporated inside FamiWare, it will be used each time an application decides to use the extension of FamiWare with INTER-TRUST. There are other actions, such as an action to connect the application components with the encryption component in the places where encryption must be ensured, that will be specific for each application. Therefore, we must provide the code that implements the security policies that are going to be considered in a certain application, the code for activating or instantiating them and the correspondence between the code and the implementation level tasks.

After implementing all of the required specific implementation tasks, FamiWare automatically generates at runtime the reconfiguration plans with the specific implementation level tasks necessary to adapt the system. In the next section, we show how FamiWare uses the implementation tasks and the correspondence between features and implementation tasks to generate, at runtime, the reconfiguration plans that need to be executed in order to reconfigure the system for security issues.

4.3. FamiWare with Security for the ITS Case Study

Now that we have integrated FamiWare and INTER-TRUST, we can generate a VSpec tree configuration of the FamiWare family for the different nodes of our ITS case study (*i.e.*, the vehicle sensors and devices, the road sensors, the RSU devices and the central server). Focusing on security, each device in the ITS case study has different necessities with regard to its security level. They also have different necessities regarding the reconfiguration of the security at that device.

Figure 8 shows a configuration of the VSpec tree for our case study. It is not complete so as to focus only on showing the different security levels that can be configured for each node of the network. Concretely, we have provided details for three different kinds of devices. An example of the first type of device is the weather sensors at the left-hand side of Figure 8. These sensors are highly constrained devices installed in the roads in order to obtain precise information about the weather conditions. This information allows the adaptation of the speed to the changes in the weather. The information provided by these sensors is critical from a security point of view, since an intruder could change it to indicate a sunny day when really it is raining, causing an accident if the speed limit is increased in bad weather conditions. This could be avoided by cyphering the information sent by the weather sensors as is shown in Figure 8, where the security/DataSecurity/cypher features have been selected. However, no reconfiguration support of any kind is going to be provided for these devices, and thus, the FamiWare services for reconfiguration were not selected.

An example of the second type of device is the traffic cameras. These are also constrained devices where the security is critical. First, only authorized users will receive the information transmitted by the cameras. Second, all of the information transmitted by the cameras must be encrypted. In order to deploy the required security level in the traffic cameras, the access control/authorization and the DataSecurity/cypher features need to be selected for this specific configuration. Note that only four features from the security variability model defined in Figure 6 have been selected to configure the security level of the traffic cameras. The application developer has considered that in order to avoid the consumption of battery and memory in traffic cameras, these devices will be deployed with an initial security configuration that will not be changed at runtime. Thus, among the services deployed for these devices, the INTER-TRUST modules will not be deployed. However, the difference with the weather sensors is that, in the traffic cameras, other resources, such as the battery, can be monitored (by the monitoring service) and could be reconfigured at runtime (by the reconfiguration service) according to the contextual information received (by the context awareness service).

Finally, an example of the third type of device is the vehicle's OBU. These devices have more resources, and thus, a security solution that can be dynamically reconfigured at runtime can be deployed with fewer limitations on the consumed resources. Thus, in Figure 8, we can see how, under the services of the OBUs, all of the modules provided by INTER-TRUST in order to negotiate and reconfigure the security level of applications will be deployed. Moreover, a large set of security concerns, including authentication, authorization, encryption of data and privacy, are deployed in these nodes.

5. DYNAMIC RECONFIGURATION OF SECURITY IN FAMIWARE

FamiWare also uses the VSpec tree described in the previous sections at runtime to drive the reconfiguration of the middleware with self-adaptation purposes. Concretely, FamiWare uses a models@run.time [23] approach to generate and to interpret the reconfiguration plans with the tasks to execute for switching from a configuration of the middleware to a new configuration in response to context changes. By using a models@run.time approach, the changes are performed at the architectural level, and thus, the architectural configurations running in each network device, and at all times, are correct according to the global network and system restrictions. Other approaches implement reconfiguration mechanisms at the code level [24,25], but the main drawback is that everything is hard-coded, so the correspondence between the reconfigured code and the models used at design time is lost. This makes it impossible to ensure that each middleware instance will enter a valid state after a reconfiguration is executed in the whole system.

Following an autonomic computing approach [26], the monitoring, context awareness and reconfiguration services, described in the previous section, collaborate in order to reconfigure the middleware at runtime. In Figure 9, we focus on the reconfiguration of security, but a similar process is followed to reconfigure any other functionality in the WSN. The only difference is the use of the INTER-TRUST modules that have been specifically introduced to manage security in FamiWare.

Concretely, Figure 9 represents the process followed when the SDS-based election feature is selected from the VSpec tree shown in Figure 5. Steps A and B represent the context observation and context analysis tasks, while Steps C, D and E represent the dynamic reconfiguration of security in FamiWare. In Step A, we see how the reconfiguration of security can occur due to the monitoring of the security contexts previously modeled or by a renegotiation of the security policies using the INTER-TRUST negotiation service. When this occurs, the changes need to be analyzed (Step B) to decide whether or not the middleware self-adaptation is required. Using the models@runtime approach provided by FamiWare (C1), this self-adaptation (self-protection when we talk about security) implies: (1) the generation of a new reconfiguration plan (Step D) using the models available at runtime (C1 and C2), the contextual information provided by Step A and the SDS file generated by the policy interpreter module in Step B; and (2) the execution of the reconfiguration plan (Step E) to adapt the middleware services (security in this case).

The reconfiguration plans are calculated by applying a difference operator between two configurations/products, the active one (C2) and the new configuration to be deployed (C3). The plans are carried out to guarantee that the system functions after a context variation, and the main goal is to change from a VSpec tree configuration to another valid configuration. The tasks are performed both at the VSpec tree level and at the implementation level with a semantic correspondence between them. As described in the previous section, at the implementation level, a task could be: modifying the parameter of a component, connecting two components, removing a component or sleeping sensor devices, *etc.* At the VSpec tree level, every task is interpreted as either a feature addition, a feature removal or a feature modification. The changes can be coarse grained (e.g., adding a new middleware component as, for instance, a new authentication component) or fine grained (e.g., modifying a parameter of an existing middleware component, such as, for instance, changing the certificate entity used by an existing authentication component).

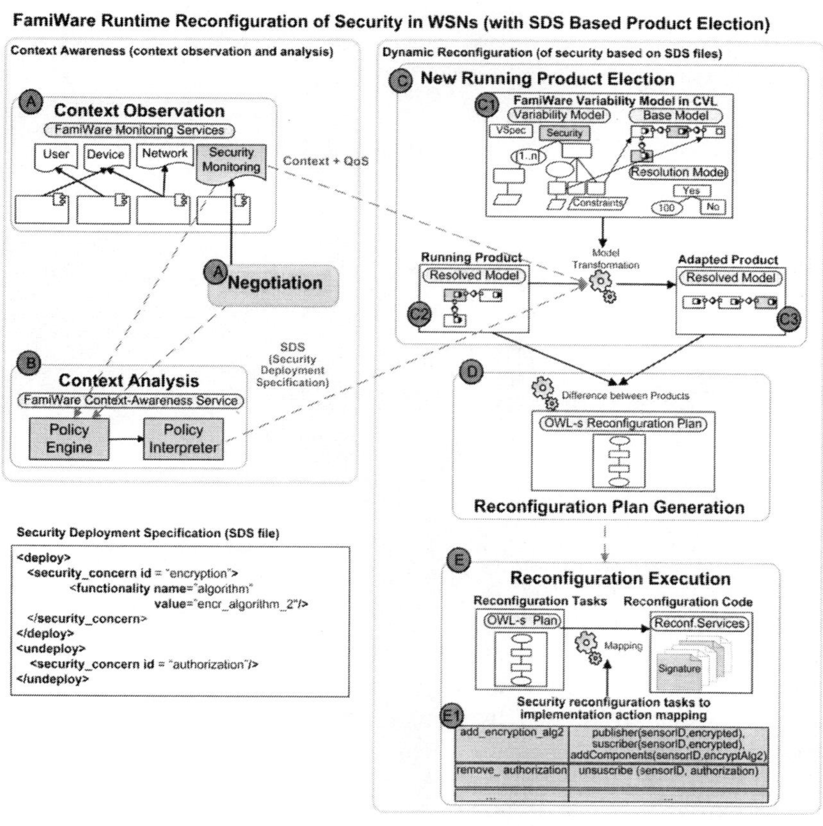

Figure 9. FamiWare runtime reconfiguration of security.

FamiWare allows the generation of plans both at design and at runtime. In the case of the security reconfiguration, the plans need to be generated at runtime, because the information to go from the existing configuration to a new one is dynamically generated by the policy engine + policy interpreter modules. This information is in the SDS file that is used to transform the VSpec tree of the running product to the VSpec tree of the adapted produced, as shown in Figure 9.

In that SDS file, we see how the output of the policy interpreter indicates that the encryption concern using the encr_algorith_2need to be deployed and that the authorization concern must be undeployed from the system. Using this SDS file, the running product (C2) and the variability model (C1), FamiWare automatically obtains the configuration that represents the adapted product (C3). In this configuration, the DataSecurity and cypher features will be selected and the authorization feature unselected. Then, using the difference operator, a reconfiguration plan with two tasks (one to add the encryption components and another to remove the authorization ones in all sensors of the WSN) is automatically generated:

```
<Sequence>
  <AtomicProcess ID="AddService">
    <Input ID="AllSensors"/>
    <Input ID="Encryption"/>
  <AtomicProcess ID="RemoveService">
    <Input ID="AllSensors"/>
    <Input ID="Authorization"/>
</Sequence>
```

As we explained in the previous section, this plan is interpreted by the FamiWare reconfiguration service deployed in each WSN node. The specific reconfiguration code for reconfiguring the security policies must be provided, so we use the mapping defined in the previous section between the security reconfiguration tasks and the specific implementation actions (label E1 in Figure 9), which needs to be executed for the system reconfiguration. In the mapping of the example, we can observe that for adding an encryption algorithm to a sensor, as FamiWare follows a publish/subscribe architecture, we need to execute the following actions: (1) from now on, this sensor will publish its data under the encrypted topic; (2) this sensor is now subscribed to the encrypted topic (*i.e.*, it will receive encrypted messages); (3) the components that implement the selected encryption algorithm have to be added to this sensor. In a similar way, to remove the authorization, this sensor has to unsubscribe from this topic.

6. EVALUATION

The evaluation of the approach presented in this paper has been organized into two subsections. In the first subsection, some quantitative experimental results are shown. Then, in the second subsection, we discuss both the benefits and the main limitations of our approach.

6.1. Experimental Results

The goal of the experiments presented is three-fold: (1) they quantify the memory footprint used and the spent battery by the new security self-adaptation (*i.e.*, self-protection) mechanism of FamiWare; (2) they evaluate the scalability of FamiWare, regarding the number of security policies and reconfiguration plans; and (3) they calculate the overhead produced by the security self-adaptation mechanism, measuring the latency, since the new configuration model has been generated and the reconfiguration has been accomplished. For the experiments, we have measured the FamiWare implementation realized for MICAz motes with TinyOS [27].

- Resource consumption: Since FamiWare is composed of a variable set of services (including the new security and reconfiguration modules, among others), the memory usage on a sensor depends on how many services are in the current configuration and on the size of each one. Furthermore, since any TinyOS application is compiled together with the OS in one image, we also have to include the OS as part of our measures. Specifically, to perform the reconfiguration mechanism of the security policies, apart from a WSN routing protocol to transmit data, the minimal instantiation must have one security monitoring service, the context awareness service and the reconfiguration service plus a simple security module. Therefore, in this case, using, for instance, the Drip [28] routing protocol, the memory footprint is greatly reduced. Considering that typical MICAz nodes have 4 KB of RAM, this instantiation consumes 51% (21,340 bytes) of RAM. On the other hand, the maximum architectural configuration for reconfiguring the middleware, considering all of the security monitoring services to observe different kinds of security contexts, the microkernel, Drip protocol and a simple application (123 bytes) and several security modules, uses 80% (3400 bytes) of the memory, so there is still 20% free for other services and applications. Note that this maximum configuration, which includes services to monitor the global status of a part of the network, is only instantiated in cluster heads, which are usually device sensors with higher capacities than the regular MICAz.

With respect to the evaluation of the energy consumption of the security dynamic reconfiguration of FamiWare, we have measured the battery waste of such process in a WSN. The battery capacity of the MICAz motes is 2000 mAh and lasts about a year. We have simulated several reconfigurations of a 20-node network using security reconfiguration plans similar to the ones defined in the previous sections (Sections 4 and 5), when the energy of the nodes is around 30% (600 mAh) on average. We compare these results with the energy expenditure of the same system, but without reconfiguration and, therefore, maintaining the same security policy defined at design time. The average of the network residual energy starts to decrease when the reconfiguration is being performed (simulation interval from 0 to 1000 s). This is due to the extra cost of operations for reconfiguring the nodes and the sink. For instance, they must be in the "active" state with the radio chip "on", and they must access their flash memory. During the reconfiguration, the remaining energy average drops to 59,984 mAh for the first 1000 s simulated, while in the system without reconfiguration, it only drops to 599,911 mAh. There is a difference of 0.071 mAh, which is the price of reconfiguration in terms of energy, but after it finishes, the energy expenditure is more or less equal for the reconfigured system as for the static one (depending on the security policy installed, since some of them consume more energy than others). This amount of spent battery (0.071 mAh) is similar to the amount that the same WSN wastes when the sensors are monitoring several issues (e.g., temperature and humidity) with 1 s of frequency, and they spread this sensed information throughout the network. Therefore, we consider that this extra waste of energy is acceptable for these long-lived systems, considering that they will provide more flexible security levels than static systems.

Summarizing, the resource consumption of FamiWare for the security dynamic reconfiguration is satisfactory and highly suited to the memory and battery constraints of real sensors.

- Scalability: We assess the scalability performance in terms of the number of plans that an ordinary node can load in its flash memory. This will be related to the number of different security policies that can be managed, since in the security context, we need a different plan to change from one security policy to another one. MICAz motes have 512 KB of flash, 384 KB of which are used for the routing protocol and for loading the binary code of the new images [27]. Therefore, if we reserve 98 KB for loading the plans, there is still 30 KB free, enough for applications or sensed data. The size average for the plans (after unnecessary tags of the OWL-S file are removed) is about 500 bytes, so there is enough space to load 200 plans, this is the upper limit of the number of plans. On the other hand, in an ordinary node (not sink or cluster-head), we can reduce the space reserved for plans, since these nodes do not need this high number of plans loaded inside the memory. The time needed to search a plan in the flash memory increases when there are numerous plans loaded. As Figure 10 depicts, the increase of time is linear with respect to the number of plans. Finally, the maximum time needed to find a plan, if we consider 200 plans loaded, is 1298 ms, an insignificant delay, compared with other sensor tasks.

Considering that we need a plan to switch from one security policy to another, FamiWare works well with 14 different security policies, since, to quickly manage 14 policies, we need to have 14 × 14 = 196 plans loaded in the sensors. In the case that we need to manage more policies, we can load new plans on the fly. This increases the reconfiguration time due to communication delay, as we see in the latency evaluation.

- Overhead and Latency: As for the negotiation, the generation of the new configuration and the selection of the plan are all done outside the sensor devices; we just measure the interval time from when the reconfiguration service receives the plan to change the security policy until the reconfiguration has been completed in all of the nodes. The average time taken by the reconfiguration service in different security policy modifications is 570 m. The longest time in this service corresponds to the execution of the tasks that comprise the plans. This is because many tasks implies the reconfiguration of several remote nodes. Therefore, this entails having to send several messages to different nodes, and the time this takes depends on the routing protocol used. For instance, using Drip protocol in one hop (without intermediate nodes), it takes 8 ms, but if we consider three intermediate nodes, this time increases up to 6.43 s. It is worth noting that if we increase the number of sensor nodes, but they are still one hop from the sink, the time needed to reconfigure them does not increase significantly. Nevertheless, if the sink has to use some intermediate nodes to reconfigure other nodes, the time greatly increases, but it depends on the structure of the network and the routing protocol used, as in other WSN platforms. Therefore, the overhead produced to reconfigure the system is insignificant when compared with the necessary time to send a packet.

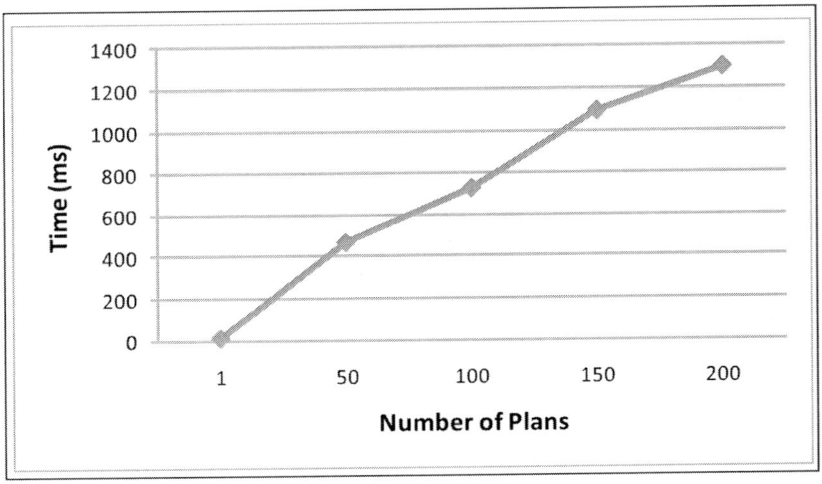

Figure 10. Time that FamiWare takes for finding and reading a plan.

6.2. Discussion

Analyzing the overhead, latency and resource consumption, we can conclude that the adaptation of the security policies to the contexts has an extra cost (especially in tiny devices like sensors), but in return, we obtain secure applications optimized for the different situations. This implies that the applications will be more secure and will consume only the necessary resources. Nevertheless, in cases of critical situations, such as emergency scenarios, where the time response has to be very short at a certain moment, the reconfiguration must be postponed if and when the instantiating of the adapted security policies would not be strictly necessary for the correct functioning of the system. Therefore, in these cases, we must avoid the reconfiguration if it is only being done for the optimization of the security, especially if this reconfiguration implies having to self-adapt many devices and takes a long time. We therefore need to realize some trade-offs, taking into account the expected response time, the security needs and the reconfiguration overhead, to decide when the reconfiguration must be carried out.

On the other hand, using models@run.time, the product running in each device corresponds to the resolved model that satisfies each context. Then, as we have generated them, respecting the constraints between the variable elements, we can ensure that the global system enters a valid state after the reconfiguration. Furthermore, as FamiWare defines the dependencies between the different sensor nodes as constraints between clones in the variability model, the running product after the reconfiguration maintains the compatibility between the devices. Nonetheless, the only models that are loaded in the sensors and sent over the network are the reconfiguration plans that consist of small files (around 500 bytes), twenty times smaller in size than the variability model.

Therefore, we ensure the correct functioning of the system after the reconfiguration if the security policies and reconfiguration task implementations respect the models. However, thus far, we have not monitored this reconfiguration process, nor have we observed that the system works as expected at runtime. This is done in INTER-TRUST using both active testing and fuzz testing techniques [19], where these modules verify, before and during the operational phases, that the negotiated contracts are respected and that the security policies are respected in dynamic systems. However, these modules of INTER-TRUST have not been integrated into FamiWare, yet.

Regarding the necessity of the evolution of these kinds of systems where new technologies are appearing everyday, using the models@run.time and the SPL paradigm to model the security, the addition of new security policies (or the modification of existing ones) is very easy. As the security concerns are modeled in the VSpec tree shown in Figure 6, the new security policy has to be added to this composite variability, as well as the correspondent modules in the base model. FamiWare will then be in charge of reconfiguring the system for this new security policy in the same way as it did for the rest of the policies. Therefore, we can say that our process is prepared for the evolution. However, this evolution is not totally automatic, because the implementation code of the new policy and the specific reconfiguration actions has to be manually provided to FamiWare.

Finally, we have tested our reconfiguration process for the ITS case study that has been previously used to validate both INTER-TRUST and FamiWare. However, INTER-TRUST has also been tested in an e-voting scenario, and FamiWare has been implemented for ambient assisted-living applications. Therefore, it would be useful to also test our security reconfiguration process in both case studies.

7. RELATED WORK

The Related Work section is organized into two subsections. The first subsection discusses those approaches that provide support for the dynamic adaptation of security, in comparison with the INTER-TRUST framework. The second subsection discusses the main characteristics of existing DSPL approaches, in comparison with the reconfiguration provided by FamiWare middleware.

7.1. Dynamic Adaptation of Security

There are a lot of approaches that try to deal with runtime adaptation of security. For instance, in [29], the authors present a policy-based approach for automating the integration of security mechanisms into Java-based business applications. They use security@runtime, a Domain-Specific Language (DSL), for the specification of security configurations based on authorization, obligation and reaction policies. The approach presented in this paper, in contrast, is suitable for using security policies specified in any model (e.g., OrBAC—Organization-Based Access Control), since the mapping between the policies and the security functionalities is made at an abstract level of the variability model. Another difference with the approach presented in this paper is that we separate the monitoring of changes in the application and the integration of the security functionality following the MAPE-K (Monitoring, Analysis, Plan, Execution—Knowledge) loop, while they integrate the security functionalities in the same monitoring events. Moreover, they implement the security rules in separate classes, but this code is application dependent, while in our approach, the security rules do not need to be hard-coded, improving the evolution of the policies.

In [30], the authors use policy-based security profiles to make logical and knowledge-based decisions within open service environments, and it uses a layered holistic model [31] to describe security (*i.e.*, security requirements are defined using security profiles that describe the interlinking of security policies to instances of services). However, in our approach, the security policies are decoupled from the specific knowledge of the application and from the implementation of the security functionality in various aspects, and this improves the reuse of both the security policies and the security functionalities.

Model-driven security (MDS) is often used to adapt security dynamically following different approaches: UMLSec (UML for secure systems) [32], SecureUML (access control language based on UML) [33], OpenPMF (Open Policy Management Framework) [34,35], SECTET (framework for model-driven security) [36], *etc.* For instance, in [37], models@runtime is used to keep an architectural model synchronized with a policy, but this approach only supports access control policies and not any kind of security functionality, as does our security adaptation service. This is a general limitation in many security policy-based approaches, because they are primarily focused only on access control or authorization concerns [34,35,38,39] or they are focused only on a specific domain, such as mobile cloud [40] or service-oriented architecture (SOA) [34–36,38].

There are also some generic approaches for reconfiguration at runtime that do not only concentrate on security concerns. In [41], Gamez *et al.* propose a reconfiguration mechanism that switches between different architectural configurations at run-time. The configurations are based on the specialization of feature models, and the reconfiguration plans are automatically generated from the differences between them. They propagate changes in configurations at the architectural level instead of the direct aspect implementation, as we do.

7.2. Dynamic Software Product Lines

There are several approaches that propose the use of DSPL to provide reconfiguration behavior. However, they provide heavyweight solutions for general purpose platforms or applications, which are not applicable to resource-constrained pervasive devices, such as mobile devices or sensors [42,43]. In [42], the application of model-driven development and middleware technologies for supporting the development and operation of dynamic adaptive systems is presented. In [43], a feature-oriented approach to dynamically develop reconfigurable core assets is proposed. As in FamiWare [44], the authors of both proposals argue that dynamic systems can be considered as a product family line in which variabilities are bounded at runtime instead of at design time. Nevertheless, they do not provide a complete implementation of a reconfiguration process for WSN applications as FamiWare does.

More specifically, in pervasive systems, in [45], it is suggested that autonomic behavior can be achieved by using variability models at runtime. In that approach, feature models are used to guide the reconfiguration process in a similar way, as we do with CVL, but this solution is not proven for tiny devices, like sensors [44]. In [46], it is argued that DSPL can solve many challenges of the reconfiguration in intensive embedded systems, and the authors propose a runtime mechanism to change structural variability at execution time. They analyze the possible application of this proposal in four different application domains, but they do not propose a generic framework to provide context adaptation as FamiWare does. Finally, in [47], CANDEL (Context As dyNamic proDuct Line) is presented, a product line-based dynamic context management for pervasive applications. This is a generic context information representation framework that considers the context as a dynamic product line composed of context primitives specified as part of a metamodel. The development of this metamodel is presented as the first step towards a model-driven approach for the development of context-aware adaptive pervasive applications, but no model-driven process has been completed nor implemented.

Summarizing, all of these proposals promote, as does ours, the use of models to automatically adapt the systems to the context whilst maintaining a correspondence between models and code after the reconfiguration and ensuring that the reconfiguration is done safely. However, none of them provide an implementation that works for real sensor devices. Therefore, they cannot be used for security policy reconfiguration in WSNs, as we do with FamiWare.

8. CONCLUSIONS AND FUTURE WORK

By combining the contributions of the FamiWare family of middleware and the contributions of the INTER-TRUST security framework, in this paper, we have presented an approach for building self-protected WSNs. Our approach allows us to: (1) build secure applications that run on heterogeneous nodes with limited capabilities; (2) dynamically negotiate the security level of the network nodes; (3) monitor the changes in the context in order to be able to react to those changes; (4) reconfigure the security level of the applications at runtime, and (5) perform the previous tasks using an efficient solution that is able to run in WSNs.

As regards future work, the first step will be, as we already mentioned in the Discussion section, the integration with the active and fuzz testing modules of INTER-TRUST to be able to verify at runtime that the system works as expected after the reconfiguration. These modules will be integrated in FamiWare as provided services in the same way as we have done in the work presented here, with the negotiation, security monitoring and security analysis modules.

Moreover, as FamiWare proposes an SPL approach to manage the large number of different devices in today's systems, as part of our future work, we plan to extend our approach to other Internet of Things devices, not only certain kinds of sensors, but also mobile devices. This extension for other kinds of devices will then be validated using other case studies, the e-voting and ambient assisted living applications.

ACKNOWLEDGMENTS

Research funded by the European FP7-317731 project, by the Spanish TIN2012-34840, P09-TIC-5231, P12-TIC1814 projects and by the International Campus of Excellence Andalucía TECH.

AUTHOR CONTRIBUTIONS

FamiWare is the doctoral thesis of Nadia Gamez, supervised by Lidia Fuentes. All the authors participates in the INTER-TRUST project. The integration of INTER-TRUST into FamiWare was conceived and designed by Mónica Pinto, Nadia Gámez and Lidia Fuentes and performed by Mercedes Amor, José Miguel Horcas and Inmaculada Ayala. The experiments were designed by Nadia Gámez, and carried out and analyzed by Nadia Gámez, Inmaculada Ayala, Mercedes Amor and José Miguel Horcas. Mónica Pinto, Nadia Gámez and Lidia Fuentes wrote this manuscript, which has been reviewed by Mercedes Amor, Inmaculada Ayala y José Miguel Horcas. The work was conceived within research projects led by Lidia Fuentes, whose main tasks are led by Mercedes Amor and Mónica Pinto.

CONFLICTS OF INTEREST

The authors declare no conflicts of interest.

REFERENCES

1. Khan, M.K.; Xiang, Y.; Horng, S.J.; Chen, H.H. Trust, Security, and Privacy in Next-Generation Wireless Sensor Networks. *Int. J. Distrib. Sens. Netw.* **2013**, *2013*, 2.

2. Shi, W.; Gong, P. A New User Authentication Protocol for Wireless Sensor Networks Using Elliptic Curves Cryptography. *Int. J. Distrib. Sens. Netw.* **2013**, *2013*.

3. Wang, J.; Ni, D.; Li, K. RFID-Based Vehicle Positioning and Its Applications in Connected Vehicles. *Sensors* **2014**, *14*, 4225–4238.

4. Cho, E.J.; Hong, C.S.; Lee, S.; Jeon, S. A Partially Distributed Intrusion Detection System for Wireless Sensor Networks. *Sensors* **2013**, *13*, 15863–15879.

5. Qiu, Y.; Zhou, J.; Baek, J.; Lopez, J. Authentication and Key Establishment in Dynamic Wireless Sensor Networks.*Sensors* **2010**, *10*, 3718–3731.

6. Hallsteinsen, S.; Hinchey, M.; Park, S.; Schmid, K. Dynamic Software Product Lines. *Comput. J.* **2008**, *41*, 93–95.

7. Hinchey, M.; Park, S.; Schmid, K. Building Dynamic Software Product Lines. *Comput. J.* **2012**, *45*, 22–26.

8. INTER-TRUST European Project. ICT-2011.1.4 Trustworthy ICT Contract n. 317731. Available online: http://www.inter-trust.eu (accessed on 3 March 2015).

9. Gámez, N.; Fuentes, L. FamiWare: A family of event-based middleware for ambient intelligence. *Pers. Ubiquitous Comput.* **2011**, *15*, 329–339.

10. Haugen, O.; Wąsowski, A.; Czarnecki, K. CVL: Common Variability Language. *SPLC'12* **2012**, *2*, 266–267.

11. OMG, Object Management Group. Available online: http:// www. omg.org (accessed on 3 March 2015).

12. OCL 2014, Object Constraint Language, 2.4, 2014. Available online: http://www.omg.org/spec/OCL/2.4/ (accessed on 3 March 2015).

13. OMG 2008, Data Distribution Service for Real-Time Systems, v1.2., 2008. Available online: http://doi.ieeecomputersociety. org/10. 1109/MS.2014.24 (accessed on 3 March 2015).

14. Fayad, M.E.; Schmidt, D.C.; Johnson, R.E. *Building Application Frameworks: Object-Oriented Foundations of Framework Design*; John Wiley & Sons, Inc.: New York, NY, USA, 1999.

15. Autrel, F.; Cuppens, F.; Cuppens-Boulahia, N.; Coma, C. MotOrBAC 2: a security policy tool. Proceedings of the 3rd Conference on Security in Network Architectures and Information Systems (SAR-SSI 2008), Loctudy, France, 13–17 October 2008; pp. 273–288.

16. ETSI: European Telecommunications Standards Institue. *ETSI TS 102 637-1 V1.1.1 (2010-09)* Technical Specification.**2010**.

17. Butun, I.; Morgera, S.; Sankar, R. A Survey of Intrusion Detection Systems in Wireless Sensor Networks. *IEEE Commun. Surv. Tutor.* **2014**, *16*, 266–282.

18. Lee, K.; Kang, K.C.; Lee, J. Concepts and Guidelines of Feature Modeling for Product Line Software Engineering. Proceedings of the 7th International Conference on Software Reuse: Methods, Techniques, and Tools, Austin, TX, USA, 15–19 April 2002; Springer-Verlag: London, UK ICSR-7. ; pp. 62–77.

19. Toumi, K.; Cavalli, A.; Aouadi, M; Ayed, S.; Cuppens, N.; Cuppens, F.; Idrees, S.; Mallouli, W.; Montes de Oca, E.; Fuentes, L.; *et al.* INTER-TRUST Approach and Framework Specification (First Version), 2013. Available online: http://www.inter-trust.eu (accessed on 3 March 2015).

20. Kiczales, G.; Lamping, J.; Mendhekar, A.; Maeda, C.; Lopes, C.; Loingtier, J.M.; Irwin, J. Aspect-Oriented Programming. In *ECOOP'97—Object-Oriented Programming*; Springer-Verlag: Berlin-Heidelberg, Germany, 1997; Volume 1241, LNCS; pp. 220–242.

21. Gámez, N. FamiWare: An Autonomic Middleware Family for Ambient Intelligence. Ph.D. Thesis, University of Málaga, MÃąlaga, Andalusia, Spain, 2011.

22. OWL-S, Semantic Markup for Web Services. Available online: http://www.w3.org/Submission/OWL-S/ (accessed on 3th March 2015).

23. Blair, G.; Bencomo, N.; France, R. Models@ run.time. *Comput. J.* **2009**, *42*, 22–27.

24. Liu, T.; Martonosi, M. Impala: A Middleware System for Managing Autonomic, Parallel Sensor Systems. Proceedings of the Ninth ACM SIGPLAN Symposium on Principles and Practice of Parallel Programming, San Diego, CA, USA, 11–13 June 2003; ACM: New York, NY, USA; pp. 107–118.

25. Heinzelman, W.; Murphy, A.; Carvalho, H.; Perillo, M. Middleware to support sensor network applications. *IEEE Netw.* **2004**, *18*, 6–14.

26. *IBM 2005, Autonomic Computing White Paper — An Architectural Blueprint for Autonomic Computing, 2005*; IBM Corp: New York, NY, USA, 2005.

27. Gámez, N.; Fuentes, L.; Aragüez, M.A. Autonomic Computing Driven by Feature Models and Architecture in FamiWare. Proceedings of the 5th European Conference on Software Architecture (ECSA 2011), Essen, Germany, 13–16 September 2011; pp. 164–179.

28. Levis, P.; Patel, N.; Culler, D.; Shenker, S. Trickle: A self-regulating algorithm for code propagation and maintenance in wireless sensor networks. Proceedings of the First USENIX/ACM Symposium on Networked Systems Design and Implementation, San Francisco, CA, USA, 29–31 March 2004; pp. 15–28.

29. Elrakaiby, Y.; Amrani, M.; Le Traon, Y. Security@Runtime: A Flexible MDE Approach to Enforce Fine-grained Security Policies. *Eng. Secur. Softw. Syst.* **2014**, *8364*, 19–34.

30. Tan, J.J.; Poslad, S. Dynamic security reconfiguration for the semantic web. *Eng. Appl. Artif. Intell.* **2004**, *17*, 783–797.

31. Tan, J.J.; Poslad, S.; Titkov, L. A semantic approach to harmonizing security models for open services. *Appl. Artif. Intell.* **2006**, *20*, 353–379.

32. Jrjens, J. *Secure Systems Development with UML*; Springer-Verlag: Berlin/Heidelberg, Germany, 2010.

33. Basin, D.; Doser, J.; Lodderstedt, T. Model Driven Security: From UML Models to Access Control Infrastructures. ACM Trans. *Softw. Eng. Methodol.* **2006**, *15*, 39–91.

34. Lang, U. OpenPMF SCaaS: Authorization as a Service for Cloud & SOA Applications. Proceedings of the 2nd IEEE International Conference on Cloud Computing Technology and Science, Indianapolis, IN, USA, 30 November–3 December 2010; pp. 634–643.

35. Lang, U. Cloud & SOA Application Security as a Service. In *ISSE 2010 Securing Electronic Business Processes*; Vieweg+Teubner Verlag: Berlin, Germany, 2011; pp. 61–71.

36. Katt, B.; Gander, M.; Breu, R.; Felderer, M. Enhancing Model Driven Security through Pattern Refinement Techniques. In *Formal Methods for Components and Objects*; Lecture Notes in Computer Science; Beckert, B., Damiani, F., de Boer, F., Bonsangue, M., Eds.; Springer-Verlan: Berlin/Heidelberg, Germany, 2013; Volume 7542, pp. 169–183.

37. Morin, B.; Mouelhi, T.; Fleurey, F.; Traon, Y.L.; Barais, O.; Jézéquel, J.M. Security-Driven Model-Based Dynamic Adaptation, Proceedings of the IEEE/ACM International Conference on Automated Software Engineering, Antwerp, Belgium, 20–24 September 2010; pp. 205–214.

38. Dong, W. Dynamic Reconfiguration Method for Web Service Based on Policy. Proceedings of the 2008 International Symposium on Electronic Commerce and Security, Guangzhou City, China, 3–5 August 2008; pp. 61–65.

39. Gheorghe, G.; Crispo, B.; Carbone, R.; Desmet, L.; Joosen, W. Deploy Adjust Readjust: Supporting Dynamic Reconfiguration of Policy

Enforcement. In *Middleware 2011*; Springer-Verlag: Berlin/Heidelberg, Germany, 2011; Volume 7049, pp. 350–369.

40. Cho, H.S.; Hwang, S.M. Mobile Cloud Policy Decision Management for MDS. In *Convergence and Hybrid Information Technology*; Springer-Verlag: Berlin/Heidelberg, Germany, 2012; Volume 7425, pp. 645–649.

41. Gámez, N.; Fuentes, L. Software Product Line Evolution with Cardinality-Based Feature Models. Proceedings of the 12th International Conference on Software Reuse (ICSR'2011), Pohang, Korea, 13–17 June 2011; pp. 102–118.

42. Bencomo, N.; Sawyer, P.; Blair, G.; Grace, P. Dynamically Adaptive Systems are Product Lines too: Using Model-Driven Techniques to Capture Dynamic Variability of Adaptive Systems. Proceedings of the 2nd International Workshop on Dynamic Software Product Lines, Limerick, Ireland, 8–12 September 2008.

43. Lee, J.; Kang, K. A. Feature-Oriented Approach to Developing Dynamically Reconfigurable Products in Product Line Engineering. Proceedings of the 10th International Software Product Line Conference, Baltimore, MD, USA, 21–24 August 2006.

44. Gámez, N.; Fuentes, L.; Troya, J. Self-Adaptation of Mobile Systems with Dynamic Software Product Lines. *IEEE Softw.* **2014**.

45. Cetina, C.; Giner, P.; Fons, J.; Pelechano, V. Autonomic Computing through Reuse of Variability Models at Runtime: The Case of Smart Homes. *IEEE Comput.* **2009**, *42*, 37–43.

46. Bosch, J.; Capilla, R. Dynamic Variability in Software-Intensive Embedded System Families. *IEEE Comput.* **2012**, *45*, 28–35.

47. Jaroucheh, Z.; Liu, X; Smith, S. CANDEL: Product Line Baed Dynamic Context Management for Pervasive Applications. Proceedings of the International Conference on Complex, Intelligent and Software Intensive System, Krakow, Poland, 15–18 February 2010.

10

OPEN HARDWARE: A ROLE TO PLAY IN WIRELESS SENSOR NETWORKS?

Roy Fisher [1,2,*], Lehlogonolo Ledwaba [3], Gerhard Hancke [2] and Carel Kruger [1,3]

[1] Advanced Sensor Networks Research Group, University of Pretoria, Lynwood Road, Pretoria 0083, South Africa

[2] Department of Computer Science, City University of Hong Kong, 83 Tat Chee Avenue, Kowloon Tong, Hong Kong

[3] Meraka Institute, Council for Scientific and Industrial Research South Africa, Meiring Naude Road, Pretoria 0184, South Africa

ABSTRACT

The concept of the Internet of Things is rapidly becoming a reality, with many applications being deployed within industrial and consumer sectors. At the 'thing' level—devices and inter-device network communication—the core technical building blocks are generally the same as those found in wireless sensor network implementations. For the Internet of Things to continue growing, we need more plentiful resources for building intelligent devices and sensor networks. Unfortunately, current commercial devices, e.g., sensor nodes and network gateways, tend to be expensive and proprietary, which presents a barrier to entry and arguably slows down further development. There are, however, an increasing number of open embedded platforms available and also a wide selection of off-the-shelf components that can quickly and easily be built into device and network gateway solutions. The question is whether these solutions measure up to built-for-purpose devices. In the paper, we provide a comparison of existing built-for-purpose devices against open source devices. For comparison, we have also designed and rapidly prototyped a sensor node based on off-the-shelf components. We show that these devices compare favorably to built-for-purpose devices in terms of performance, power and cost. Using open platforms and off-the-shelf components would allow more developers to build intelligent devices and sensor networks, which could result in a better overall development ecosystem, lower barriers to entry and rapid growth in the number of IoT applications.

KEYWORDS

Internet of Things; industrial networks; machine to machine; smart grid; wireless sensor network

1. INTRODUCTION

The Internet of Things is a concept in which more devices (objects/things) will be communicating across the Internet than human users [1]. These connected and communicating devices will enable wide ranging applications. Already, applications from building automation [2] to smart-agriculture [3] have been deployed and tested, with much research being done into the next big thing within this new ecosystem [4–7].

Along with the increase in the interest and research devoted to the Internet of Things, a new development strategy has become more prominent in the Internet of Things ecosystem [8]. The development of both hardware and software using an 'open-source' approach has gained much attention [9]. Open source is the idea that all design and implementation documents get released in an open forum; it is, however, still possible for the original creator to control the licensing rights of the original design. Open source can extend to both hardware and software. In software, this includes all lines of code; in hardware, this typically includes documents describing the implementation. Open source hardware is a more recent development than open source software. Open source not only allows designers to create their own devices and systems from existing proven designs, but enables companies to sell complete open source systems to consumers. This allows for the consumer to design and improve the system as they see fit, but giving them a concrete standard and working starting point, typically known as a COTS (commercial off-the-shelf) system. One important aspect that needs to be considered is the use of open source technologies to implement the solutions required for the Internet of Things. Open source technologies and systems have a number of advantages and disadvantages [10]. One of the major advantages is that the system is generally designed with the wider community in mind; this means that compatibility issues are less likely. This model does lend itself to abuse, and all open source software and hardware should be thoroughly tested before being applied in highly sensitive applications.

Currently, the Internet of Things is an attractive idea for industrial applications [11,12]. Allowing data acquisition across multiple facets of an industrial life cycle could greatly increase efficiency, as well as allow the implementer a greater deal of insight and control over their industrial activities [13]. A possible application scenario would involve a great deal of embedded devices in a single environment. The major concerns for industrial implementations of the Internet of Things are: capability, reliability and cost [14]. With the increasing following and emerging standards for wireless communication, learning from and improving upon old technologies will allow for the creation of platforms that will still be developmentally relevant in the future. This will allow for developer communities to build and establish code bases and standards on which future platforms would be based. This would go a long way in being able to establish hardware and software standards for the Internet of Things.

However, the availability and cost of intelligent 'things' are arguably inhibitors of very large-scale adoption. More inexpensive technology could be used, such as radio frequency identification (RFID), but apart from identification and data capture, this does not truly realize an intelligent and connected device. Implementations of the Internet of Things designed for a specific application also appear to have seen little movement forward in terms of design updates. In this paper, we focus on the possibility that the development of supporting technologies for the Internet of Things could be based on generic and open source technologies that can be molded to suit the required task. It is our view that this approach would more rapidly increase the number of devices within the Internet of Things. Raspberry Pi is an example of such a device that could potentially be used in a number of applications [15,16]. However, a comparative study has not yet been completed as to the full capabilities of Raspberry Pi or similar platforms, when compared to 'built-for-purpose' Internet of Things devices, such as commercial sensor network nodes and gateways. The ease with which new technologies, such as Raspberry Pi, Beaglebone and the Arduino board, can be expanded, through modular expansion boards, and built upon will allow for an easy to implement and robust application using these devices. The application of modern, modular technologies will allow for an increase in the flexibility of the Internet of Things [17]. These technologies are aiming first and foremost to be generic, allowing for a more standardized approach to the Internet of Things. Designing generic technologies, which follow clearly predefined standards, will allow for a large degree of interoperability between all technologies. Industrial and consumer electronics have also reached a stage where the designer can easily and quickly use COTS modules and reference designs to build advanced embedded devices, which arguably provide comparable or improved performance in terms of power consumption, communication and processing resources when compared to built-for-purpose devices.

We believe that these advantages weigh heavily in the case for the application of open source technology to the Internet of Things. In a building automation example, the developer could deploy a relatively inexpensive open source COTS base system (Raspberry Pi) with a suitable modular communication capability (ZigBee) and easily design his own application to run on this system. As new communication technologies become available through the years, a simple upgrade to the communication module of the application is possible without completely redesigning the application or changing the already deployed base system [18]. This makes the application more flexible, cost effective and capable over the long term. Keeping these advantages in mind, we have decided to examine the possibility of taking open source Internet of Things technologies and applying them to wireless sensor networks [19].

Due to these advantages, one of the most appealing examples of the combination of the two separate paradigms is in the application of WSN and Internet of Things technology to the application of smart grids [20]. The focus is not on creating a hardened technology for important tasks within a factory environment. The idea is to introduce the Internet of Things technology gradually by initially using it in non-critical systems within a wireless sensor network. Using this design approach, we believe the open source concept is future proof, uses an advanced open source operating system, is inexpensive and as good as built-for-purpose components.

In order to provide some credibility for our belief, we tested built-for-purpose devices against both open source components and a COTS rapidly prototyped device that was created, in conjunction with the CSIR (Council for Scientific and Industrial Research in South Africa (http://www.csir.co.za/), using up to date components. We also created a testbed allowing us to benchmark our open source choices against the industry built-for-purpose devices that are currently available. The final aim of our paper is to provide an overview of all of the different technologies that are available and, through our testbed, give a blueprint to how these technologies can be used together. The paper details a cross-section of the most promising options that are available for Internet of Things applications. Additionally, these options are compared to some of the devices that were part of the first generation of sensor network technologies. This was done to act as a gauge of the progress that has been achieved through the development of the technologies related to the Internet of Things. We also created a device from COTS components, in order to provide a simple guide to better show the possibilities available when implementing and designing Internet of Things applications, especially when creating it for specific application areas.

2. ARCHITECTURE

A generic implementation of the Internet of Things can be seen in Figure 1. The generic implementation shows a simple view of an embedded Internet of Things implementation. The Internet of Things device cloud is a traditional wireless sensor network. In this simple view, the gateway device would communicate with one of the many platform options that are available through standard technologies allowing for Internet connectivity [21].

In the image, two distinct networking zones can be seen. The division between these two networking zones exist because of the requirements of the devices that inhabit these areas of the architecture. One method of allowing these different networking zones to communicate is to make use of a gateway routing device.

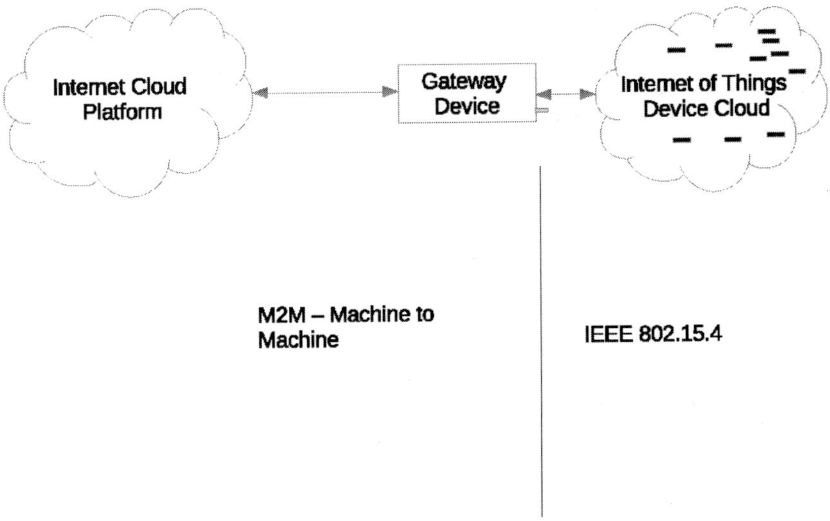

Figure 1. Internet of Things generic network implementation.

2.1. Internet of Things Device Cloud

The device cloud is the implementation of a traditional wireless sensor network (WSN). These wireless sensors typically have two major functions: they allow for the sensing of the environment, and sometimes, they allow for the WSN to act on the environment (through an actuator).

The Internet of Things device cloud will consist of nodes that have at least the following capabilities:

- Self organizing, allowing for the determination of routes, as well as handling failed nodes;
- Protection of the data sent between the nodes and gateway devices;
- Communication capabilities for a large number of devices.

Some of the embedded protocols that these devices make use of are proprietary, and the gateway devices that will be used will need to be able to communicate across the chosen set of protocols.

2.2. Gateway

This section of the architecture for the Internet of Things typically involves the use of a gateway device to communicate and, in some cases, translate the device cloud communications onto the wider Internet, thereby allowing any Internet-capable device to communicate with the sensor and actuator device cloud [22,23]. Many options exist for the gateway devices. These gateway devices will have the capabilities of allowing for the translation between the Internet of Things device cloud and the cloud-based platform. These gateway devices will typically spend most of their time performing tasks similar to a routing device. The application of security features, such as encryption, to the data from the device cloud to the platform will also fall upon this gateway device. These gateway devices must be able to communicate with conventional network devices, as well as the low-power network devices embedded within the device cloud.

2.3. Cloud-Based Platforms

The back-end server devices are generally deployed in two separate ways: either the servers are deployed in-house or the developers use a platform to manage the server devices. These platforms are typically deployed on the cloud. This makes the deployment more elastic and, in some circumstances, more reliable. Some of the cloud platforms that are currently deployed are gaining more and more acceptance worldwide and becoming the standard method for creating Internet of Things applications. The two approaches both have their advantages and disadvantages. A number of articles have been written regarding this proposed structure [24]. A number of platform options exist, namely Axeda, Bugswarm, Carriots, Thingspeak and Xively. The purpose of this area is to collect, analyze and present the data in a method that the end user can both understand and make use of. This area is therefore primarily concerned with the presentation of the data to the client.

3. NODES

Many different approaches have arisen to creating an Internet of Things application. Some of the current ranges of embedded enabling devices for the Internet of Things device cloud are introduced below [25].

We noticed that the divide between open source and closed source seems to be that the newer devices follow an open source approach, whereas the older versions follow a closed source approach. Waspmote, Lotus, Sprouts, Firefly and TelosB are open source. MICAz and Imote2 are closed source devices. A few of the devices mentioned as open source are not completely open source due to external problems and the inability to use the advantages of being an open source device. A good example of this is the TelosB device: the ability to advance or add attachments to the device is limited because of the design. Therefore, although being an open source device, due to external complications, it does not gain the advantages of other open source devices and is therefore very similar to a closed source device.

Waspmote Pro

Waspmote Pro is a new open source, wireless sensing node from cooking hacks by Libelium, improving on their initial Waspmote node [26]. Waspmote Pro can be used in various industries, such as gas and events monitoring, smart metering, agriculture and radiation detection, owing to a number of sensor boards available for use with the node. This node falls under our built-for-purpose devices, as it has been designed for a specific task and contains specific sensors and equipment.

Lotus

Lotus is an advanced WSN node built around the ARM Cortex M3, 32-bit processor [27]. It is fitted with a 802.15.4 compliant on-board radio associated with the ZigBee protocol. The Lotus is well suited for applications requiring acoustic, video, vibration and other high-speed sensor data, condition-based maintenance, industrial monitoring and analysis and seismic and vibration monitoring.

Sprouts

The Sprouts node is a developing node from Queens University [28]. Developed as part of industry-related research, the node has attracted attention in the oil and gas mining, steel production and power grid monitoring sectors as an event monitoring node. This is one of the nodes that falls under our open source options with many expansion capabilities.

Firefly 2.2

Developed at Carnegie Mellon University, the Firefly 2.2 is a real-time sensor networking node designed for use in industrial control and automation, smart home monitoring, inventory and personnel tracking and hazardous environment monitoring, as well as general embedded system education [29]. This is one of the nodes that falls under our open source options with many expansion capabilities.

Older Options

The newer options that have been discussed above are not the only options that are available. The other options to be discussed below are: TelosB, MICAz and Imote2. There are a number of other options that could be discussed, and a complete list of node options are available [25].

Some other examples are WeC, Rene, Dot Mica, Spec, Cricket, EyesIFX, Tmote Sky, Shimmer, Stargate, Sun SPOT, IRIS, NetBridge, BTNode, V-Link and a few others [25]. Many of these examples are closed source. These devices were all created before 2009 and can be traced back to the designs used within TelosB, MICAz or Imote2. We have therefore decided to make use of these three devices as representative of the older devices that are available. The oldest versions are well represented by TelosB and MICAz (two of the first iterations of the Internet of Things), and through the years, the upgrades can all be represented through Imote2. These decisions were made after carefully studying the components used to create the different iterations of all of these available devices.

Rapidly Prototyping New Devices

Alternative options to the ones listed do exist. One of the major options that we have is to create our own device by making use of COTS components [30]. This allows us to personalize our devices to perfectly suit our requirements. The advantage gained from being able to create a node that specifically addresses your requirements must be done with the other devices in mind. The huge range of technologies used to create the Internet of Things often means that, within a single application, a large range of technologies are used. For example the 802.15.4 protocol is not the only protocol being implemented within this section of the architecture. Industrial production lines using the more mature WiFi protocol also exist [31]. WiFi technology is based on the 802.11 protocol and has been in use in both residential and industrial applications for many years. The 802.11 protocol, although capable of being implemented within the device cloud section of the Internet of Things, was originally designed for implementation within devices with high resource allowance (power, throughput and large memory space) [32]. The standard has evolved, and the current implementation, 802.1 lac, focuses on high data throughput and multiple access [33]; both are less important for the Internet of Things device cloud, where a large number of devices communicate irregularly. Using a device such as Raspberry Pi and a plugin XBEE or WirelessHart device will enable a powerful range of capabilities for an embedded sensor. These are a few of the considerations that must be taken into account when developing your own node.

Creation of your own node requires only a basic understanding of electronics, communication interfaces and embedded programming. To illustrate the feasibility of rapidly prototyping a self-built solution, we created a simple node to test alongside the nodes discussed above. The node that we prototyped is based on a DiZiC module (http://dizic.com/), with our own simple 'break-out' board and additional communication and power circuitry. The node has full communication capabilities based on the IEEE 802.15.4/ZigBee protocol stacks. The developed node can be seen in Figure 2, and the functional diagram is given in Figure 3. This shows the simple design of the implemented node and the range of communication options available to the node. The main chip was chosen to ensure a wide range of compatibility with as many of the available Internet of Things operating systems as possible. The other components were chosen to ensure that the device remained as cost effective as possible, but could still perform the required operations. An advantage of making use of such a design is to ensure that the device meets one's individual requirements and is able to meet the goals of one's application, but remain easy to both replace and upgrade. This ensures that the entire application remains robust, inexpensive and flexible, all of which ensure a future-proof design. This node that we have created is capable of performing all of the functions that the other devices can achieve and when compared against a number of the available options and at a much lower cost, even when considering that we were ordering small quantities.

3.1. Device Comparison

The primary comparison of the devices will look at their basic functionality and capability [28]. The initial testing will look at the power consumption of each of the devices.

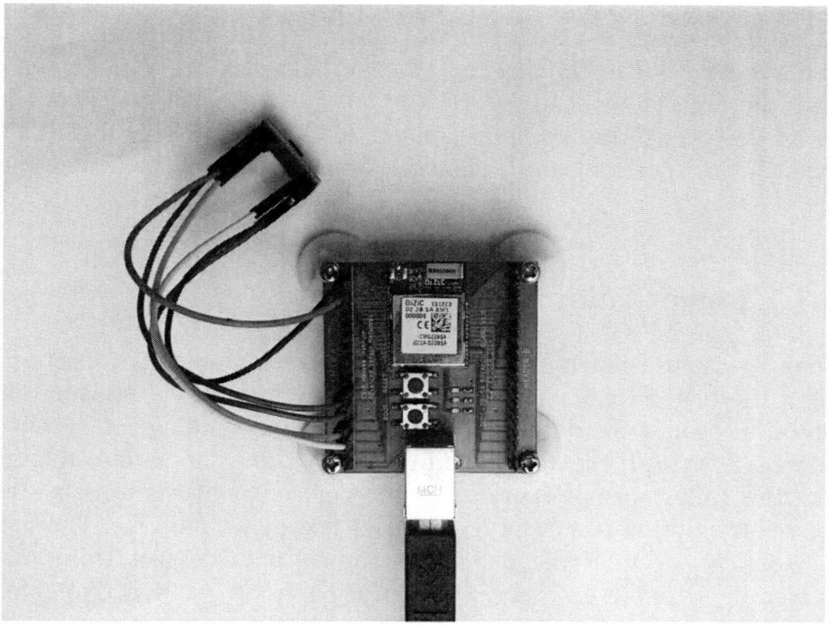

Figure 2. The rapidly prototyped Council for Scientific and Industrial Research (CSIR) Internet of Things node.

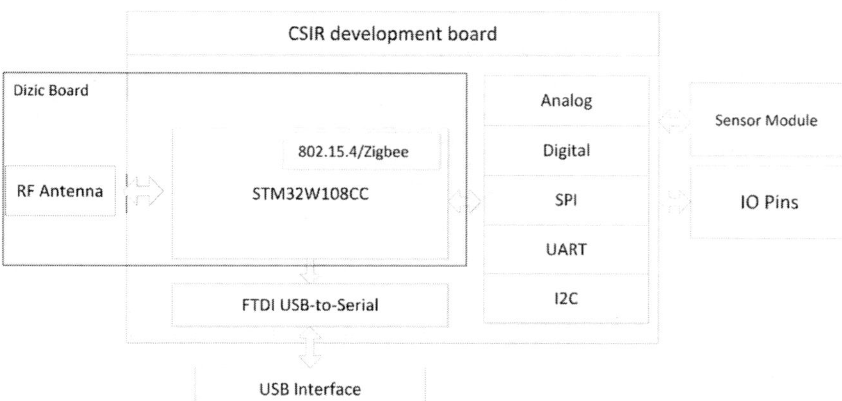

Figure 3. The CSIR Internet of Things node functional diagram.

The list of devices above can be broken down into two separate categories: traditional wireless sensor network nodes and non-traditional (devices that were not originally created to perform the role of a wireless sensor network node). Raspberry Pi and Beaglebone are the non-traditional WSN nodes, as they have not been implemented with the limitations of the WSN in mind. These devices have not been developed to use the minimal amount of power, *etc.* This means that these devices perform very well in certain aspects of our benchmarking.

3.1.1. Power Comparison

The two graphs seen in Figures 4 and 5 show the power consumption of all of the tested devices. These graphs show the current consumption (in mAh) against the duty cycle (% of time spent sending and receiving information) for these individual Internet of Things nodes. For the time that the nodes are not operational, the device spends its time in sleep mode, attempting to conserve as much energy as possible. The most economical of all of the devices is the Sprouts device. This device successfully achieves less than 180 μAh in full active mode (continuously awake). Some of the other economical devices are Waspmote, which achieves a maximum of 15 mAh in full active mode. The older devices that have been compared here all perform very similarly. These are TelosB and the MICAz; they both achieve around 25 mAh for full awake operation. The Firefly node achieves a similar current requirement of 24.8 mAh for continuous operation. These devices are followed closely by Imote2 and Lotus platform, each achieving 44 mAh and 66 mAh, respectively, for continuous operation. Using these graphs and by knowing the power limitations of the application being designed for, a suitable node can be chosen beforehand. This allows for adequate planning and can help prepare a suitable budget when deciding on which node to use for the Internet of Things application.

As can be seen from the graphs shown above, our rapidly prototyped COTS node has only slightly higher power requirements, mostly due to the more powerful processor that is included on the board, than a cluster of current sensor node options. The node achieves 31 mAh when operating at 100% and 400 nAh when in sleep mode. The sleep mode current is high for the device, but the operational current is similar to what is required for the other nodes.

The power consumption of the non-standard nodes (by non-standard, we mean any device that is not originally created for use within the Internet of Things/WSNs) is significantly higher. The sleep and non-sleep of these nodes is defined differently from the sleep and non-sleep of the traditional nodes. A traditional WSN node sleep typically involves a state in which most power consuming hardware is off and waiting for a wake command, either from the onboard processor or another device in the network. The sleep for Raspberry Pi and Beaglebone is defined as a state of rest, where little to no processing is occurring (an expansion board for Raspberry Pi can be added, which will enable a WSN node, like sleep mode). The traditional sleep (for WSN without the expansion board) can be achieved, but it is very difficult to wake the device, and it takes a significant length of time longer than the more traditional nodes. As can be seen from Figure 5, Raspberry Pi consumes 44 mAh for continuous operation, and Beaglebone consumes 450 mAh for continuous operation.

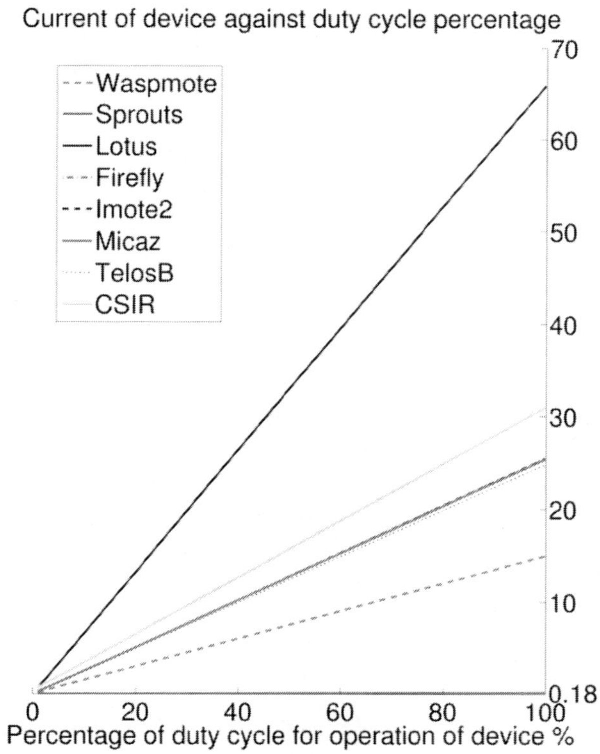

Figure 4. Wireless sensor nodes' current consumption against duty cycle.

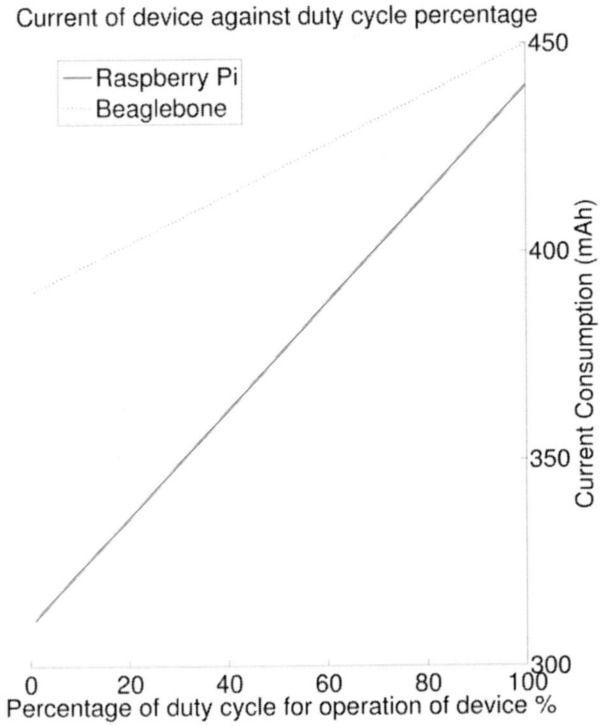

Figure 5. Non-traditional nodes' current consumption against duty cycle.

3.1.2. Capabilities Comparison

In the current world of computing, more cost typically gives you greater capability. A comparison was completed between the cost of the nodes and the processing capability that is received for the cost. Processing capability is not the only performance-based metric that can be used. This metric was chosen, as it is generally the one most linked to the cost of the node and potentially the one with the most impact on actual performance.

The cost of the Firefly device is currently not known, as it is still under development. The Sprouts device is open source, and although the cost of the completed device is not available, the bill of materials cost was used, as a capable individual could in theory create one of these devices themselves. Surprisingly, in this comparison (Figure 6) the most powerful devices are also the least expensive. The Sprouts node is the least expensive complete WSN node, with our own rapid prototype device coming in as being one of the least expensive available, as well. Although both Sprouts and the CSIR-made device are low on processing and power, depending on the application, this should not be a drawback from the implementation within a wireless sensor network. The low power requirement will allow for long-term deployments within Internet of Things applications, such as those within embedded environment monitoring applications, and with a careful design, the low processing power will not negatively affect the performance of the application.

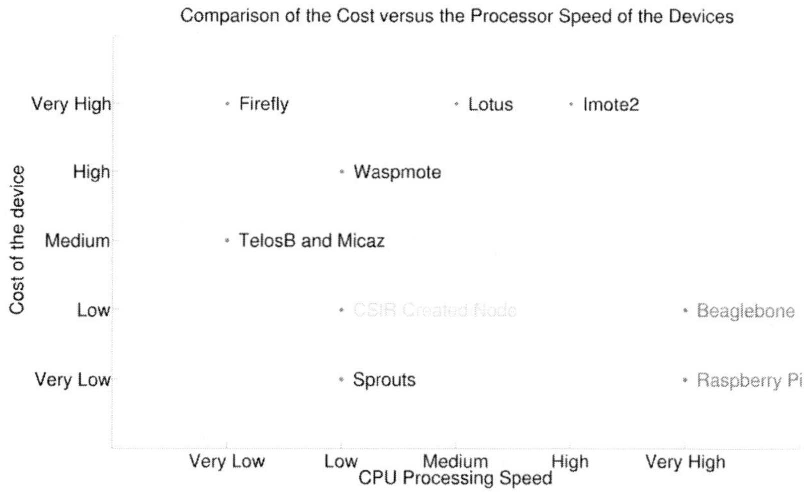

Figure 6. Comparison of processing speed against cost for WSN nodes (commercial off-the-shelf (COTS) and open platforms indicated in green and red, respectively).

The next two least expensive are Raspberry Pi and Beaglebone. Both of these devices will require a XBee module (or similar) for wireless communication. These modules cost about $30–$40. This will not change the ranking of these two devices on the scale. This graph shows that some of the least powerful devices are in fact the most expensive. The Lotus node and the Imote2 node both cost $300, a massive sum to be paying for a WSN node. The built-for-purpose devices end up being expensive, possibly due to the ease of plug and play capability. The additional cost for Raspberry Pi and Beaglebone is getting the devices to communicate successfully with the network through the general purpose input output (GPIO) serial ports. Due to the open source nature, many tutorials and online examples exist to assist with this problem.

3.1.3. Features Comparison

This comparison looks specifically at the hardware and software capabilities of the individual devices (costs shown in the table are rough guides, as large discounts can be achieved when ordering large quantities).

As can be seen, the features of the devices differ greatly. One important feature is the operating system of the node chosen. The operating system has an impact on the stability, capability and security of the device. A well-developed and continuously-updated operating system can allow for new powerful features to be deployed to the nodes without making any hardware changes. Raspberry Pi and Beaglebone both have many options available, and many of these options are well supported in terms of continuous updates. TinyOS and Contiki are both open source, with active communities driving development. NanoRK is an operating system created by Carnegie Melon University for use on their WSN node. Mote Runner is a set of tools to aid the running of the node from IBM [34]. Table 1 shows that the features of the devices can differ greatly. However, it also shows that the current communication choice for Internet of Things embedded nodes is the ZigBee protocol. This is due to the large number of devices that support this communication standard, as can be seen from Table 1. A number of the devices also include built-in onboard sensors that can be used for sensing applications, giving a reduction of the cost for certain applications when these onboard sensors are used. Some of the devices also include a range of GPIO ports available on the devices. These devices allow for advanced features to be added to the application. As can be seen, a range of features is available to the nodes.

It can be seen from the table that our node, although being a rapid implementation, has similar capabilities to the other nodes tested. This shows that the use of the circuits available and a simple working knowledge of general electronic skills can allow one to create powerful and inexpensive devices. Our node has full 32-bit capability, is coded in a standard language (C++) and has a range of expansion capabilities through the generic I/O pins. The node is also significantly less expensive than many of the built-for-purpose devices.

Table 1. Table showing features of the WSN nodes. GPIO, general purpose input output.

Node	OS	CPU Architecture (bits)	CPU Speed (MHz)	Network Standard	No. of Sensors	Language	I/O Pins	Release Date	Cost ($)
Waspmote	None	8	14.75	Various (Depends on Implementation)	2	C/C++	21	2013	210
Sprouts	Linux	32/8	32	ZigBee, BLE	2	C	0	2011	20
Lotus	Mote Runner/ TinyOS	32	10–100	ZigBee	0	nesC	0	2011	300
Firefly	NanoRk	8	7.3	ZigBee	0	C	20	2010	Un-known
Imote2	TinyOS	32	13–416	ZigBee	0	nesC	12	2006	300
MICAz	TinyOS	8	16	ZigBee	0	nesC	4	2004	100
TelosB	TinyOS or Contiki	16	16	ZigBee	3	nesC	16	2005	120
CSIR Node	Contiki	32	24	802.15.4/ ZigBee	0	C/C++	17	2012	80
Raspberry Pi	Raspbian, Android, Many more	32	700	None, but expandable [1]	0	Any for Linux OS	0 GPIO and UART, I^2C SPI	2012	35
Beagle-bone	Angstrom, Raspbian, Many more	32	720	None, but expandable [2]	0	Any for Linux OS	65 GPIO and UART, I^2C SPI	2012	89

[1] Raspberry Pi expansion board (http://elinux.org/RPi_Expansion_Boards); [2] Beaglebone expansion boards (http://elinux.org/BeagleBone_Community).

Table 1 shows as many of the capabilities of each of the individual devices. Some of the additional expansion capabilities are included in footnotes at the bottom of the page, and these contain links to developer web pages, showing a large list of expansion capabilities available for Raspberry Pi and Beaglebone.

4. GATEWAY

When considering the performance capabilities of gateway devices in an industrial implementation, the most important characteristic is communication performance. To prevent data bottlenecks and slow transmission when measuring performance, three key areas must be considered: reliability, throughput and response time [12].

Reliability measures the confidence in a transmitted packet being received by the cloud platform. For a gateway in an industrial application, high reliability is required. This enables the application implementer to be sure that transmitted data are received by the cloud platform. High reliability will ensure that data transfer rates are not negatively affected by retransmission of data across a low resource network. Typically, networks based on the 802.15.4 protocol stack do not implement a transmission control protocol [35]. This means that reliable transmission is not guaranteed by the communication protocol. A high reliability will allow for a greater degree of confidence that transmitted data have reached the destination. Reliability must also take into account the order in which the packets are received. In a time-sensitive implementation, which has limited resources, any additional processing must be avoided [36].

Throughput is a measure of the amount of data that the system can process per second. The gateway implementation will need to take data from one communication technology and retransmit on the other end. This will typically involve receiving data from a 6LoWPAN network and re-transmitting it on a normal IPv6 or IPv4 network. The higher the throughput of the device, the better for the implementation. Another measure of the throughput is the latency of the device. Latency can also be called response time. This is a measure of the time it takes for another device to respond. A low latency is preferred, as this means that the gateway is operating more efficiently.

Due to the nature of the information that will be transferred across the Internet by Internet of Things applications, the additional process of adding security to the communication needs to be benchmarked [37]. The three major requirements for the Internet of Things security is confidentiality, integrity and authentication [38,39]. These three requirements are known as the primitive security objectives. The processing requirements of these primitive objectives are extensive and will be a good benchmarking tool for the devices, but these devices will also be required to implement these objectives in communication in order to secure the communication [40]. Confidentiality is the ability to hide the data that are transmitted. Integrity is the ability to ensure that the data that are transmitted are the same as the data that are received. Authentication is used to ensure that the individuals communicating are the people that are expected to be part of the communication. Using all three of these objectives allows us to ensure data security when communicating.

To complete the benchmarking of the possible devices, two approaches could be used: built-in toolkits for benchmarking; or the creation of scripts that perform the required benchmarking.

4.1. Open Source Implementation

Raspberry Pi has often been labeled as a 'hobby' device [15]; this leaves questions as to whether or not this device is capable of operating as a gateway or in any time-/data-sensitive application. Due to the increasing community interest in Raspberry Pi, the device has undergone a large amount of testing and benchmarking already. Another device to consider for implementation is Beaglebone. This device has been benchmarked, but is not considered as an easy alternative to Raspberry Pi. The focus of the paper is to supply a current state-of-the-art for the implementation of an Internet of Things application. Beaglebone, although being a very capable device, has not seen the interest and development from third party individuals that Raspberry Pi has enjoyed. Beaglebone has a less well-supported operating system, known as Angstrom Linux. Although the operating system on both devices can be changed, the paper looks at the default design of the devices, and Raspberry Pi's Raspbian operating system has more support and software packages due to its Debian heritage.

During the benchmarking of the devices, it was shown that Beaglebone and Raspberry Pi perform on similar levels. Beaglebone lacks some of the easy expansion capabilities of Raspberry Pi. The best example is the ability of Raspberry Pi through a $40 board to be able to use Arduino shields. In order to easily use a Xbee module, a $6 board is required [41]. It was therefore decided that Raspberry Pi would represent the open source implementation of a gateway device.

4.2. Proprietary Implementation

The proprietary implementation of the gateway devices involves the use of the Digi Connectport devices acting as the gateway for the embedded Internet of Things sensors to the wider Internet. As discussed previously, the two devices that are tested are the Connectport X2 and the Connectport X4. Both of these devices have LAN and ZigBee networking capabilities. Unlike Raspberry Pi, these devices have a very limited set of customization options available. They both run the Digi-implemented operating software that has a specific focus on operating requirements [42,43].

4.3. Experimental Details

A set of experiments were conducted that showed the capabilities of Raspberry Pi with regards to the requirements for an open source gateway device, namely reliability, throughput and security. The aim of these experiments was to determine whether or not Raspberry Pi is a device capable of acting as a gateway for the Internet of Things in an industrial application. Tests were done to show each of the requirements and to measure Raspberry Pi's capabilities with regards to these requirements. Each of the tests will be discussed in the sections below. Due to the interest in Raspberry Pi, many benchmarking tests have already been completed. These benchmarking tests (specifically the OpenSSL tests) use the built-in benchmarking capabilities provided within the OpenSSL development kit. Although being a good performance comparison between the different devices available, it does not provide a real-world comparison of the devices. The benchmarking tools allow for uniform data to be worked upon in a structured and controlled loop only utilizing high-speed sections of onboard hardware (RAM). This is the ideal operating environment for benchmarking, but does not provide a real-world view of operational capabilities. A real-world implementation involves more operational calls to external input output devices and interrupts from other devices.

This is why when completing the implementation, these external features were simulated within the code, allowing for a more realistic view of the real-world operational capabilities of the device.

4.3.1. Reliability Test

The first set of tests were done to confirm the reliability of Raspberry Pi. Reliability testing was done using a widely-known network benchmarking tool. This is used to ensure that the packets sent and received are reliably delivered. It should be noted that our definition of reliability is somewhat limited. We have decided to only focus on the successful delivery of packets between the receiver and transmitter. This was to provide us with a solid starting point for comparison against the two security protocols tested later in the section. Although being a limited approach, it provides us with a good basis for testing the security protocols. The reliability could be affected by nodes moving in or out of the network range, as well as other possible faults that could arise. The test consisted of a laptop connecting to Raspberry Pi via the laptops WiFi and a USB WiFi receiver on Raspberry Pi. Raspberry Pi hosted a WiFi access point and was also connected to a desktop computer via the onboard Ethernet capabilities. Each of these separate networks was setup in a subnet. Raspberry Pi had IP forwarding enabled, switching the packets between the Ethernet and WiFi networks. Making use of the Iperf benchmarking suite, the performance was measured [44]. When measuring the reliability of the wireless communication, the distance of the transmitter from the receiver can play an important role. The tests were run at three distance intervals, with the laptop moving away from Raspberry Pi.

The Connectport devices are designed for reliability, as they are created to be deployed in an industrial environment. The Connectport devices have a similar setup structured for the reliability testing. Due to the wired nature of their communication, distance testing was not required.

4.3.2. Throughput Tests

The throughput testing was completed using two separate tests. The raw data bandwidth was measured using the Iperf benchmarking suite for data of different sizes. Additionally, the Iperf benchmarking suite has the capability of simulating the communication of up to one hundred devices. A test was run that simulated the communication of a large number of devices across the WiFi channel to Raspberry Pi and onto the server connected via Ethernet. This simulation was run three times, and the average results were calculated. The latency of the communication was also measured by pinging the laptop and desktop computer through Raspberry Pi.

Secure Throughput Tests: Raspberry Pi Command Line

A Python script was created that could call the required OpenSSL command line tool. The Python script built the required command and then called the command using the built-in set of OpenSSL command line functions on Raspberry Pi. The command was executed on a file that was greater than 500 MB, and the time to complete the required security objective was measured. The authentication of the system is not included in the results below, as the times were consistently less than one second. This leads to unreliable results, so it is assumed that a key is used that is encrypted with the message as authentication. These tests were run a number of times, and the average throughput calculated. The tests were completed using a block size of 8,196 bytes. These tests assumed that the entire file has the required security objective applied and is then transmitted as an entire package by some underlying protocol, for example a TCP-based unencrypted communication. The underlying protocol will then break the file into packets as required. Once received, the entire file could be decrypted using a pre-shared key. For the integrity tests, the secure communication of the signature of the file was assumed to exist. The raw processing and throughput speeds are more important in terms of benchmarking Raspberry Pi. A simple solution is to transmit this information as part of the file in the first 16 bytes of the message. By doing this, no major additional processing is required; it will simply be the attachment of x (the size of the integrity proof) bytes onto the file. The aim of these tests is to determine the processing and communication time required to protect the communicated data for each of the primitive security requirements.

Secure Throughput Tests: Raspberry Pi Python Wrapper

This implementation broke up the file into equally-sized packets (1,024 bytes). Each packet then had a combination of integrity, confidentiality or authentication applied to it and was then transmitted. These tests were done to better simulate the size of data that would be received from the in-house network, as this is generally small packets from a large number of devices. The aim was to show the added overheard when dealing with the small-sized data packets that are typically going to be received and transmitted by Raspberry Pi. The M2Crypto Python wrapper was used to interface with the OpenSSL C functions [45]. An additional advantage of using a trusted and established wrapper for secure communication allowed us to confirm our results.

The next set of tests was completed using the best performing cipher suite and message digest to provide confidentiality, integrity and authentication. A system was then implemented that hashed, signed and finally encrypted each 1,024-byte message and transmitted it along the network. The message digest used was MD5, as this has proven to be a fast hashing algorithm. The signing algorithm was RSA-based, and the encryption was implemented using the AES 256 CBCalgorithm.

Secure Throughput Tests: Proprietary Implementation

Similar testing was completed for the the Digi devices. The testing allowed for secure sockets layer-based communication of data (built from the data collected from the embedded network) between the Digi device and a suitable standalone computer.

4.4. Results

4.4.1. Reliability Tests

The results for the test can be viewed in Table 2. As can be seen from these results, Raspberry Pi is capable of achieving reliable data transmission at distances up to 10 m with very few packets either dropped or arriving out of order.

The proprietary Digi Connectport X2 and Digi Connectport X4 devices have reliability and response capabilities expected of a networking device in an implementation. The devices achieve a 0% packet drop rate and a very low response time, averaging <10 ms.

Table 2. Table summarizing the results from the reliability test with a Raspberry Pi.

Distance	Result
<1 m	0% packets dropped
5 m	0% packets dropped and 1 arrived out of order
>10 m	0% packets dropped

4.4.2. Throughput Results

The results for the raw (unsecure) data throughput tests can be seen in Table 3. These data show that the larger the information being sent through Raspberry Pi, the more throughput is being achieved; the increase is due to the larger data sizes being able to stream at maximum speed for longer periods due to all of the communication, requiring setup time. The data also shows that Raspberry Pi does not have a large effect on the data flow, as the communication is limited by the WiFi standard used, namely 802.11g, which achieves a maximum of a 22-Mbps throughput [46].

Table 3. Table summarizing results from the throughput test with Raspberry Pi.

Data Size	Result
15 MB (megabytes)	11.5 Mbps (Megabits per second)
18.2 MB	15.2 Mbps
60 MB	21.5 Mbps
200 MB	22.2 Mbps

For the Iperf communication of multiple device simulation, Raspberry Pi was capable of achieving an average throughput for 100 communicating devices of 15.8 Mbps.

The ping testing was repeated 10 times, and the average time to receive a response was 80.011 ms. The minimum response time was 12.669 ms, and the maximum response time was 205.143 ms. This large swing in response time could potentially negatively affect the performance of communication through Raspberry Pi. The large swing could be due to the single core processor that is included with Raspberry Pi. The response to one of the pings is delayed, because the processor is performing some other set of tasks (other processor interrupts) before it can return to the task of responding to the request.

Secure Throughput Results: Raspberry Pi Command Line

The results for the implementation applying only confidentiality are shown in Table 4. The data throughput using the recommended encryption cipher AES with a 256-bit key achieves 3.20268 MB/s. The two fastest performing algorithms are Blowfish and AES, both achieving greater than a 3-MB/s throughput. Tests were also completed with regards to the cipher chaining method. The best performing chaining method was ECB (electronic code book); this was expected, because of the simplicity. The next best performing chaining method was CBC. AES 256 using CBC achieved a throughput of 3.279 MB/s, whereas the ECB achieved a throughput of 3.421976 MB/s.

Table 4. Table summarizing results from Python and the command line implementation of communication with confidentiality.

Confidentiality Algorithm	Throughput (MB/s)
AES128 bit key	3.34744
AES 192 bit key	3.318837
AES 256 bit key	3.20268
Blowfish	3.317878
DES	2.590775
Triple DES	1.458663
RC4	3.58629

One important consideration is RC4; it would appear that by adding I/O operations, the throughput of this algorithm is greatly reduced. The operation of RC4 is delayed due to the I/O operations that need to be performed. Whereas without the file I/O, the RC4 stream cipher could rapidly process the data due to its stream nature, complementing the continuous loop cycle of the benchmark test. It is now being hampered by the low speed of the I/O operations, but the cipher is still able to perform at the fastest rate.

The results in Table 5 show the results achieved when adding integrity to the communication with the OpenSSL command line tool. As can be seen, the results for just integrity far outperform the throughput of adding confidentiality. The best performing hashing function was MD4, achieving a throughput of 28.82474 MB/s. These results are achieved by implementing the hash on the complete file and then transmitting the hash value along with the file.

Table 5. Table summarizing results from Python and the command line implementation of communication with integrity.

Integrity Algorithm	Throughput (MB/s)
MD4	28.82474
MD5	24.85157
ripemd160	15.347
SHA	19.09429
SHA1	21.68956
SHA 224	15.2319
SHA 512	8.411452
Whirlpool	1.566124

Figure 7 shows the results for the implementation of both integrity and confidentiality on the data sent from Raspberry Pi. These results show that a data throughput rate of 2.962641 MB/s is achievable using the AES 128 encryption with MD5. Similar speeds can be achieved using AES 128 with MD4 and RC4 with SHA1. The more secure implementation of AES 256 with MD4 achieves 2.852727 MB/s. These values show that by adding I/O operations, the speed of the system is slowed drastically. It is important to note that in these tests, authentication of the communication is handled by the knowledge of the pre-shared key.

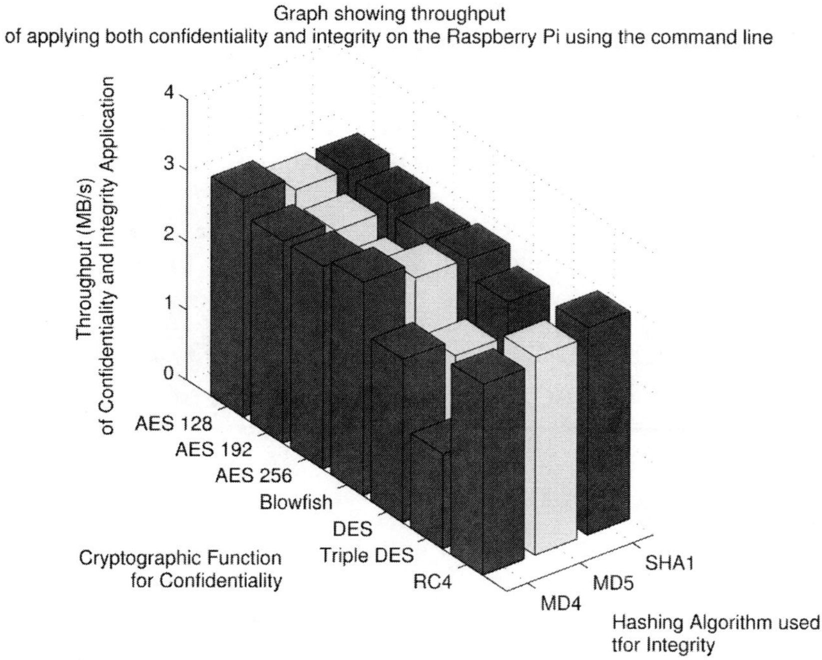

Figure 7. Throughput results when applying both confidentiality and integrity for command line application.

Secure Throughput Results: Raspberry Pi Python Wrapper

The first set of results were for the implementation of the described system passing messages of a size of 1024 bytes, encrypting, then sending the data and can be found in Table 6. As can be seen, the throughput achieved is very similar to the throughput achieved for the previous implementation using the command line. Only CBC mode was tested, because from the previous experiments, it became clear that this was the best performing mode. It is clear from these results that AES and Blowfish again achieve the fastest throughput. This is expected and confirms the results found regarding the fastest cipher [47].

The results of applying all of the primitive security objectives using the results from the previous tests are that the throughput capability of applying all of the primitive security objectives was 1.5778 MB/s.

Table 6. Table summarizing results from the Python OpenSSL wrapper implementation of communication with confidentiality.

Confidentiality Algorithm	Throughput (MB/s)
AES 256 CBC	2.221412771
Blowfish CBC	2.211554
DES CBC	1.84385782
Triple DES CBC	1.138862268
RC4	2.696817074

Tests were not implemented to confirm the capabilities of signing and hashing each protocol, as there is no point in applying each primitive security objective separately in a comparison test. One more test was completed. This test measured the throughput of Raspberry Pi when implementing confidentiality and integrity on the device. This test was done to mimic the operation using a static key for authentication. This test achieved a throughput of 1.764865 MB/s. This shows that the process of signing the information does not account for a very large percentage of the processing power of the processor. This confirms the results that where achieved in the first benchmarking results.

Secure Throughput Results: Proprietary Implementation

The results show that the X4 device performs significantly worse than the X2 device, achieving only 113 KB/s of data throughput compared to the X2 achieving 1.4 MB/s. Both of these devices perform worse than Raspberry Pi. This is surprising, since the two Digi devices are specifically designed for the application.

4.4.3. Summary

Table 7 shows a summary of the throughput results achieved in the paper.

Table 7. Summary throughput results for secure communication (CIA).

Device	Throughput	Throughput (Mb/s)
X4	113 KB/s	0.882 Mb/s
X2	1.4 MB/s	11.2 Mb/s
Raspberry Pi, Command Line	2.962 MB/s	23.696 Mb/s
Raspberry Pi, Python Wrapper	1.5778 MB/s	12.6224 Mb/s

Raspberry Pi is capable of achieving a maximum of 2.962 MB/s when applying confidentiality and integrity to the data that are transmitted by using a Python script and the OpenSSL command line. This is equivalent to 23.696 Mb/s. If a small amount is subtracted for line inconsistencies and overhead from Internet Protocols used in Internet communication, Raspberry Pi is capable of interacting with a 20-Mb/s communication line.

The Python library implementation achieves a 1.5778 MB/s or 12.6224 Mb/s on Raspberry Pi. The additional inefficiency of interacting through the library could be the reason for this decrease in raw speed. However, this still outperforms the Digi devices when implemented using the correct algorithms.

The proprietary implementation achieves a maximum of 1.4 MB/s or 11.2 Mb/s when applied on the newer X2 device and 113 KB/s or 0.882 Mb/s when applied on the older X4 device.

This shows that Raspberry Pi is capable of being used as a gateway device for the Internet of Things. Raspberry Pi successfully outperforms the Digi devices for throughput. This shows that Raspberry Pi is capable of being implemented as a gateway device in an industrial application.

Although Raspberry Pi performs very well, the simplicity of a proprietary implementation could be very appealing to a system developer.

It should also be noted that applying security to Raspberry Pi does not hugely impact the throughput capability of the device, as can be seen from the small difference in throughput with or without secure communication. Raspberry Pi is capable of acting as a gateway device, due to its stable reliability, availability and its high throughput capability Raspberry Pi, even though it has a low resource processor, is able to successfully handle all of the tasks associated with the secure communication of information between multiple embedded devices and the wider Internet platform.

Our results show that an open source, inexpensive and simple solution is available for a gateway device based within the Internet of Things.

5. PLATFORMS

Two primary examples of Internet of Things applications exist. The first option is to deploy the application using a centralized cloud-based server. The second option is to deploy a localized server within the organization and thereby have more control over the flow of their data [48]. Many of these systems already have industry partners and have begun to be deployed within some of the largest commercial institutions in the world. Using these platforms in this manner allows the application developer to focus his efforts on his application and instead leverage the power that these cloud-based systems allow for large amounts of data collection.

Typically, information is transmitted from the device to the gateway, and these data are then forwarded, with the required information (device ID, *etc.*), to the cloud platform, which collects and stores the data for access. This allows advanced data management features and analysis to be implemented easily and simply by the user and developer of the application.

The data that traverse the Internet of Things are typically going to be of extremely high importance [49]. Using the above example of a centralized server facilitating the collection of the information generated in a smart grid scenario, it can be seen that just on the basis of privacy, this collected information must be stored and transmitted using the current security standards. Examining the work in the gateway section and previously covered work, the secure transmission of the data has been covered [48].

The final stage is the storage of this information on the server side for later use. One of the most efficient ways to protect the information is to use local server-side encryption. By encrypting the data, only a user with the registered and required key can decrypt and, therefore, access the data [50]. This method allows for the protection of the information stored on the server and does not open up our system to a set of possible security flaws. Another method is to manage access to the data. By controlling who has access rights and enforcing a deletion policy if these access rights are traversed, our data can be securely protected [51].

Generally a built-for-purpose implementation will contain an in-house server that the local IT department has full control over. However, this method of deploying the platform for the application is both expensive and requires the purchasing of additional equipment.

A newer approach is to use the open source platforms that enable one's application to be built utilizing their platform as a service (PAAS). This allows the application implementers to focus rather on their own application and allows the platform to control the processing, network and storage requirements. This is generally done for a small fee or a limited free account does exist [48].

6. CONCLUSIONS

The Internet of Things is becoming a popular term in both the media and in the business world. The ability of these open source devices has been covered within the paper. This paper has shown that the devices could be used in place of a proprietary implementation. In some instances, the open source 'cheaper' devices perform better than the 'expensive' proprietary devices. This is encouraging as the inexpensive and open source devices could be a possible solution for the implementations of the Internet of Things.

One of the important devices that are benchmarked within the paper is Raspberry Pi. This device is inexpensive and easily upgradeable through simple expansion boards, allowing for varied and diverse implementations. Many of these advanced expansion boards and capabilities are due to the extensive support that Raspberry Pi has received from the development community. This has created a source of knowledge and useable program packages that have pushed Raspberry Pi into a truly multi-function device.

The ideal implementation from the results found above would be to use the Sprouts platform for the embedded device cloud and Raspberry Pi as the gateway device. By doing this, the advanced capabilities of Raspberry Pi are available, and the power saving of the Sprouts platform is utilized. This entire implementation is cost effective, as Raspberry Pi is $80 with full Xbee (ZigBee protocol capable module) capability, and each of the Sprouts platform devices will cost $20. This is much more inexpensive than any proprietary implementation that is available.

This set of results shows that an open source solution is a possibility and, in many cases, is often preferable to a built-for-purpose implementation. Using Raspberry Pi as a gateway device and either building one's own node or using the Sprouts node design, with some basic electronic skills, a powerful and cost-effective Internet of Things implementation is possible. We believe that this is the possible approach that could guide the future creation of Internet of Things devices.

The focus for the future is for the individual devices embedded within the device cloud to be directly connected to the Internet using the new Internet Protocol Version 6 addressing scheme. This is the focus for the future and is being carried out by research into areas, such as 6LoWPAN and COAP. 6LoWPAN is the protocol for converting the IPv6 protocol for use on low power personal area networks typically based on the 802.15.4 protocol. This allows for the embedded sensor or actuator to communicate directly to the cloud-based platform. In the future, this may be the preferred implementation setup.

ACKNOWLEDGEMENTS

The work described in the paper was partially supported by a grant from City University of Hong Kong (Project No. 7200375). This work is based on the research supported in part by the National Research Foundation of South Africa (Grant Reference TP1207183332) and our industry partners, Telkom, Bytes Universal Systems, Tellumat and EMC.

The grant holder acknowledges that opinions, findings and conclusions or recommendations expressed in any publication generated by the NRF-supported research are that of the author(s) and that the NRF and our industry partners accept no liability whatsoever in this regard.

AUTHOR CONTRIBUTIONS

RM Fisher and GP Hancke were responsible for the benchmarking of the individual security features across the range of the gateway devices. Together they also tested the Raspberry Pi and Beaglebone as nodes for the Internet of Things and reviewed the current approaches to platforms that are currently available. C Kruger and L Ledwaba contributed the COTS node for the research as well as data used in the benchmarking the individual Internet of Things nodes from the individual suppliers.

CONFLICTS OF INTEREST

The authors declare no conflicts of interest.

REFERENCES

1. Khan, R.; Khan, S.U.; Zaheer, R.; Khan, S. Future Internet: The Internet of Things Architecture, Possible Applications and Key Challenges. Proceedings of the 2012 10th International Conference on Frontiers of Information Technology (FIT '12), Islamabad, Pakistan, December 2012; IEEE Computer Society: Washington, DC, USA, 2012; pp. 257–260.

2. Kumar, A.; Hancke, G. An Energy-Efficient Smart Comfort Sensing System Based on the IEEE 1451 Standard for Green Buildings. *IEEE Sens. J.* **2014**, *14*, 4245–4252.

3. Kumar, A.; Hancke, G. A ZigBee-Based Animal Health Monitoring System. *IEEE Sens. J.* **2015**, *15*, 610–617.

4. Zuehlke, D. SmartFactory—Towards a factory-of-things. *Ann. Rev. Control* **2010**, *34*, 129–138.

5. Wang, M.; Zhang, G.; Zhang, C.; Zhang, J.; Li, C. An IoT-Based Appliance Control System for Smart Homes. Proceedings of the 2013 Fourth International Conference on Intelligent Control and Information Processing (ICICIP ′ 13), Beijing, China, 9–11 June 2013; IEEE: Washington, DC, USA; pp. 744–747.

6. Hancke, G.; Silva, B.; Hancke, G. The Role of Advanced Sensing in Smart Cities. *Sensors* **2013**, *13*, 393–425.

7. Kumar, A.; Kim, H.; Hancke, G. Environmental Monitoring Systems: A Review. *IEEE Sens. J.* **2013**, *13*, 1329–1339.

8. Xu, L.; He, W.; Li, S. Internet of Things in Industries: A Survey. *IEEE Trans. Ind. Inform.* **2014**.

9. Crowston, K.; Wei, K.; Howison, J.; Wiggins, A. Free/Libre Open-source Software Development: What We Know and What We Do Not Know. *ACM Comput. Surv.* **2008**, *44*, 7:1–7:35.

10. Khanjani, A.; Sulaiman, R. The Aspects of Choosing Open Source Versus Closed Source. Proceedings of the 2011 IEEE Symposium on Computers & Informatics (ISCI ′ 11), Kuala Lumpur, Malaysia, 20–23 March 2011; IEEE: Washington, DC, USA, 2011; pp. 646–649.

11. Chi, Q.; Yan, H.; Zhang, C.; Pang, Z.; Xu, L.D. A Reconfigurable Smart Sensor Interface for Industrial WSN in IoT Environment. *IEEE Trans. Ind. Inform.* **2014**, *10*, 1417–1425.

12. Sauter, T.; Lobashov, M. How to access factory floor information using internet technologies and gateways. *IEEE Trans. Ind. Inform.* **2011**, *7*, 699–712.

13. He, W.; Xu, L.D. Integration of Distributed Enterprise Applications: A Survey. *IEEE Trans. Ind. Inform.* **2014**, *10*, 35–42.

14. Cucinotta, T.; Mancina, A. A real-time service-oriented architecture for industrial automation. *IEEE Trans. Ind. Inform.***2009**, *5*, 267–277.

15. Edwards, C. Not So Humble Raspberry Pi Gets Big Ideas. *Eng. Technol.* **2013**, *8*, 30–33.

16. Kruger, C.; Hancke, G. Benchmarking Internet of Things Devices. Proceedings of the IEEE International Conference on Industrial Informatics (INDIN '14), Porto Alegre, Brazil, 27–30 July 2014.

17. Montenegro, G.; Kushalnagar, N.; Hui, J.; Culler, D. *Transmission of IPv6 Packets over IEEE 802.15. 4 Networks* Internet Proposed Standard RFC 4944. **2007**.

18. Kumar, A.; Hancke, G. Energy Efficient Environment Monitoring System Based on the IEEE 802.15.4 Standard for Low Cost Requirements. *IEEE Sens. J.* **2014**, *14*, 2557–2566.

19. Kruger, C.; Hancke, G. Implementing the Internet of Things Vision in Industrial Wireless Sensor Networks. Proceedings of the IEEE International Conference on Industrial Informatics (INDIN ' 14), Porto Alegre, Brazil, 27–30 July 2014.

20. Sauter, T.; Lobashov, M. End-to-End Communication Architecture for Smart Grids. *IEEE Trans. Ind. Electron.* **2011**, *58*, 1218–1228.

21. Castro, M.; Jara, A.J.; Skarmeta, A.F. An Analysis of M2M Platforms: Challenges and Opportunities for the Internet of Things. Proceedings of the 2012 Sixth International Conference on Innovative Mobile and Internet Services in Ubiquitous Computing, Palermo, Italy, 4–6 July 2012; IEEE: Washington, DC, USA, 2012; pp. 757–762.

22. Deru, L.; Dawans, S.; Ocaña, M.; Quoitin, B.; Bonaventure, O. Redundant Border Routers for Mission-Critical 6LoWPAN Networks. In *Real-World Wireless Sensor Networks*; Langendoen, K., Hu, W., Ferrari, F., Zimmerling, M., Mottola, L., Eds.; Springer International Publishing: Sennweid, 6330 Cham, Switzerland, 2014; Number 281 in Lecture Notes in Electrical Engineering; pp. 195–203.

23. Potter, C.; Hancke, G.; Silva, B. Machine-to-Machine: Possible Applications in Industrial Networks. Proceedings of the IEEE International Conference on Industrial Technology (ICIT '13), Cape Town, Western Cape, South Africa, 25–28 February 2013; pp. 1321–1326.

24. Piyare, R.; Lee, S.R. Towards Internet of Things (IOTS): Integration of Wireless Sensor Network to Cloud Services for Data Collection and Sharing. *Int. J. Comput. Netw. Commun.* **2013**, *5*, 59–72.

25. Akyildiz, I.F.; Vuran, M.C. *Wireless Sensor Networks*; John Wiley & Sons: Hoboken, NJ, USA, 2010.

26. Libelium. Waspmote Datasheet, 2013. Available online: http://www.libelium.com/development/waspmote/documentation/ (accessed on 30 October 2014).

27. Memsic. Lotus Datasheet, 2011. Available online: http://www.memsic.com/userfiles/files/Datasheets/WSN/6020-0705-01_A_LOTUS.pdf (accessed on 30 October 2014).

28. Kouche, A. Towards a Wireless Sensor Network Platform for the Internet Of Things: Sprouts WSN Platform. Proceedings of the 2012 IEEE International Conference on Communications (ICC '12), Ottawa, ON, Canada, 10–15 June 2012; pp. 632–636.

29. Carnegie Melon University. Firefly 2.2 Datasheet, 2010. Available online: http://www.nanork.org/attachments/77/FireFly_datasheet.pdf (accessed on 29 October 2014).

30. Kruger, A.; Abu-Mahfouz, A.; Hancke, G.P. A Rapid Prototype Sensor Network Gateway Using Commercial off-the-Shelf Components. Proceedings of the IEEE International Conference on Industrial Technology (ICIT '15), Sevilla, Spain, 17–19 March 2015.

31. Refaat, T.; Daoud, R. WiFi implementation of Wireless Networked Control Systems. Proceedings of the 2010 Seventh International Conference on Networked Sensing Systems (INSS '10), Kassel, Germany, 15–18 June 2010; pp. 145–148.

32. Varshney, U. The status and future of 802.11-based WLANs. *Computer* **2003**, *36*, 102–105.

33. Ong, E.H.; Kneckt, J.; Alanen, O.; Chang, Z.; Huovinen, T.; Nihtila, T. IEEE 802.11ac: Enhancements for Very High throughput WLANs. Proceedings of the 2011 IEEE 22nd International Symposium on Personal Indoor and Mobile Radio Communications (PIMRC '11), Toronto, ON, Canada, 11–14 September 2011; pp. 849–853.

34. IBM. IBM Research: Zurich; Computer Science. In *Mote Runner*; 2014.

35. Fisher, R.; Hancke, G. DTLS for Lightweight Secure Data Streaming in the Internet of Things. Proceedings of the 3PGCIC 9th International

Conference on P2P, Parallel, Grid, Cloud and Internet Computing, Guangzhou, China, 5–8 November 2014.

36. Xianli, Q.; Mingchao, F.; Bin, S. Coal Mine Gas Wireless Monitoring System Based on WSNs. Proceedings of the 2011 Second International Conference on Digital Manufacturing and Automation (ICDMA ′11), Zhangjiajie, Hunan, China, 5–7 August 2011; pp. 309–312.

37. Hancke, GP; Markantonakis, K; Mayes, KE. Security Challenges for User-Oriented RFID Applications within the "Internet of Things". *J. Internet Technol.* **2010**, *11*, 307–313.

38. Cheminod, M.; Durante, L.; Valenzano, A.; Member, S. Review of Security Issues in Industrial Networks. *IEEE Trans. Ind. Inform.* **2013**, *9*, 277–293.

39. Abu-Mahfouz, A.; Hancke, G. Distance Bounding: A Practical Security Solution for Real-Time Location Systems. *IEEE Trans. Ind. Inform.* **2013**, *9*, 16–27.

40. Melonyan, A.; Gampe, A.; Pontual, M. Facilitating Remote Laboratory Deployments Using a Relay Gateway Server Architecture. *IEEE Trans. Ind. Electron.* **2014**, *61*, 477–485.

41. Anonymous. Raspberry Pi as an Xbee Wireless Sensor Network Gateway. **2014**.

42. DiGi. ConnectPort X2 Datasheet, 2013. Available online: http://www.digi.com/pdf/ds_connectportx2.pdf (accessed on 18 September 2014).

43. DiGi. ConnectPort X4 Datasheet, 2013. Available online: http://www.digi.com/pdf/ds_connectportx4.pdf (accessed on 18 September 2014).

44. Anonymous. Iperf Benchmarking Suite, 2013. Available online: https://iperf.fr/ (accessed on 20 July 2014).

45. Python Community. M2Crypto: A Python Crypto and SSL Toolkit, 2013. Available online: https://pypi.python.org/pypi/M2Crypto (accessed on 18 October 2014).

46. Institute of Electrical and Electronics Engineers (IEEE). *Telecommunications and Information Local and Metropolitan Area*

Networks—Part 11: Wireless LAN Medium Access Control (MAC); IEEE Standard for Information Technology; IEEE: Washington, DC, USA, 2007.

47. Bingmann, T. Speedtest and Comparsion of Open-Source Cryptography Libraries and Compiler Flags. Available online: http://panthema.net/2008/0714-cryptography-speedtest-comparison/ (accessed on 18 October 2014).

48. Fisher, R.; Hancke, G. SSL Usage in Commercial Internet of Things Platforms. In *Radio Frequency Identification System Security*; Cryptology and Information Security Series; IOS Press: Amsterdam, The Netherlands, 2013; Volume 11, pp. 69–82.

49. Weber, R.H. Internet of Things—New security and privacy challenges. *Comput. Law. Secur. Rev.* **2010**, *26*, 23–30.

50. Arockiam, L.; Monikandan, S. Efficient Cloud Storage Confidentiality to Ensure Data Security. Proceedings of the 2014 International Conference on Computer Communication and Informatics (ICCCI '14), Coimbatore, India, 3–5 January 2014; pp. 1–5.

51. Tang, Y.; Lee, P.; Lui, J.; Perlman, R. Secure Overlay Cloud Storage with Access Control and Assured Deletion. *IEEE Trans. Dependable Secure Comput.* 2012, *9*, 903–916.

INDEX

M

N

P

Q

R

S

T

W

Z